Agronomy Gateway

THE AUTHOR

Dr. P. Balasubramanian, Teaching Assistant of Agronomy at TNAU graduated in Agriculture during 2009-2013 from ADAC and RI, TNAU, Trichy and Master Degree in Agronomy during 2013-2015 and completed Ph.D., in 2019 from AC and RI, TNAU, Madurai, He has received totally four awards at national and international levels including Young Scientist, Young Researcher and Best Doctoral Thesis. To his credit he has published five research articles, edited three books and presented more than 30 papers in the conference, seminar and summer institutes. He has cleared National Eligibility Test (ASRB, ICAR).

Dr. C. R. Chinnamuthu, Professor and Head Department of Agronomy, Tamil Nadu Agricultural University, Coimbatore. He has graduated in Agriculture in 1983, obtained Masters' Degree in Agronomy in 1985 with a specialization in weed science and completed Ph.D. in 1990 from Tamil Nadu Agricultural University, Coimbatore, India. He did his Post-Doctoral Research work at IRRI, the Philippines between 1991 and 1994. He has vast experience in the field of nanotechnology applications in agriculture. He has established a state-of-the-art Nanotechnology Research Laboratory and pioneered research work in the field of nanotechnology applications in weed management. His significant contributions are rapid detoxification of atrazine herbicide residue, management of purple nutsedge, fabrication of slow and smart release nano-formulations for season-long weed control in a rainfed and irrigated ecosystem. He has worked in various capacities in TNAU for the past 28 years as Deputy Director of Research, Professor IFFCO Chair, Principle Investigator in Agroforestry, Oilseed and Rice research projects. He guided thirteen PG and Ph.D. students and all of them were conferred with the best thesis and best paper awards for their outstanding contribution. He has received eighteen awards at national and international levels including Best Teacher and Best Researcher from TNAU and Young Scientist Award by APWSS. To his credit, he has published sixty research articles, edited three books in nanotechnology and presented more than seventy lead papers in the conference, seminar and summer institutes.

Agronomy Gateway

Dr. P. Balasubramanian

Teaching Assistant (Agronomy),
Tamil Nadu Agricultural University,
Horticultural Research Station,
Ooty, The Nilgiris, Tamil Nadu

Dr. C.R. Chinnamuthu

Professor and Head
Department of Agronomy,
Tamil Nadu Agricultural University,
Coimabtore, Tamil Nadu

Dr. D. Keisar Lourdusamy

Associate Professor and Head,
Horticultural Research Station,
Tamil Nadu Agricultural University
Ooty, The Nilgiris, Tamil Nadu

Mr. R.C. Prasanth

Ph.D., Scholar,
Department of Agronomy,
Tamil Nadu Agricultural University,
Coimbatore, Tamil Nadu

Mr. S. Thirumeninathan

Ph.D., Scholar,
Department of Agronomy,
Tamil Nadu Agricultural University,
Coimabtore, Tamil Nadu

Dr. R. Ajaykumar

Assistant Professor,
Department of Agronomy,
Vanavarayar Institute of Agriculture,
Pollachi, Tamil Nadu

Dr. H.A. Archana

Department of Agronomy,
Don Bosco College of Agriculture,
Sagayathottam, Vellore, Tamil Nadu

2021

Astral Cracker

A Division of

Astral International Pvt. Ltd.

New Delhi – 110 002

ISBN: 978-93-90435-61-6 (PB)

Published by : **Astral Cracker**
A Division of
Astral International Pvt. Ltd.
– ISO 9001:2015 Certified Company –
4736/23, Ansari Road, Darya Ganj
New Delhi-110 002
Ph. 011-43549197, 23278134
E-mail: info@astralint.com
Website: www.astralint.com

Digitally Printed at : **Replika Press Pvt. Ltd.**

Preface

The purpose of bringing out the first enlarged edition of "**Agronomy Gateway**" to present convenient and concise source of precise definitions of terms used in agricultural science to our reader engaged in research, teaching, and development of agriculture. The manuscript has might be prepared keeping in view the need of graduate and post graduate level students

This book provides detailed information about all the aspects of agronomy in agricultural practices. The different chapters in the book are compiled after analysing the information available in various sources. The book highlights ICAR at glance, agriculture current affairs, revolutions and important days related to agriculture and allied sciences, agronomy important terminologies, essential elements functions and deficiency symptoms in crops, important agronomy formulas, weed management, agronomy point to points, text books and their authors.

We are grateful to the authors for the contributions. I congratulate the Astral Publications for bringing out the publication **Agronomy Gateway.** I also thank and to the editors and the contributors for different chapters for their valuable time and efforts. We hope that this publication is well received by the students, teachers, researchers, and extension in field agronomy and other related disciplines will find this publication informative and useful.

Although finally, reader is requested to point them out so that they could be taken care of in the next edition.

P. Balasubramanian

Contents

	Preface	*v*
1.	ICAR at a Glance	1
2.	Agriculture Current Affairs – 2020-21	13
3.	Agricultural Revolutions and Important Days Related to Agriculture and Allied Sciences	25
4.	Agronomy: Important Terminologies	29
5.	Essential Elements Functions and Deficiency Symptoms in Crops	105
6.	Weed Management	115
7.	Important Agronomy Formulas	153
8.	Agronomy Point to Points	173

Chapter 1

ICAR at a Glance

The Indian Council of Agricultural Research (ICAR) is an autonomous organization under the Department of Agricultural Research and Education (DARE), Ministry of Agriculture and Farmers Welfare, Government of India.

- ☆ Formerly known as Imperial Council of Agricultural Research. It was established on **16th July 1929.**
- ☆ Union Minister of Agriculture and Farmers Welfare and President ICAR- **Shri Narendra Singh Tomar**
- ☆ Minister of State, Ministry of Agriculture and Farmers Welfare- **Shri Parshottam Rupala, Shri Kailash Choudhary.**
- ☆ Present Director General Of ICAR- **Dr. Trilochan Mohapatra**

ICAR Achievements in 2019-20

- ☆ Total 220 high yielding varieties including 101 climate-resilient, 15 multiple stress-tolerant and 17 biofortified varieties in 40 field crops released and notified for commercial cultivation.
- ☆ Breeder seed to the tune of 1,15,707 quintals of 1330 varieties covering 51 field crops was produced against the total indent of 85,752 quintals by the Central and State seed agencies.
- ☆ For strategic crops such as pulses and oilseeds, 97,346 and 28,304 quintals of certified seeds were produced through 150 and 35 seed hubs, respectively. These seed hubs are being managed by ICAR institutes and KVKs.

- 133 varieties of 35 horticultural crops were notified and released for commercial cultivation.
- Registered 13 new breeds of livestock and poultry. So far, a total of 197 breeds have been registered.
- Developed breeding protocols for 4 important food fishes and 11 ornamental finfish
    - Discovered 16 new fish/shellfish species
    - Developed 12 value added products from fish/shellfish waste for commercialization purpose.
- Land Resource Inventory (1:10000 scale) completed for 27 aspirational districts of India to enable block level land use planning and given to NITI Aayog
    - Prepared e-ATLAS on Micro-nutrients of the Indian Soils
    - Climate Resilient Villages, popularly called NICRA Villages have been upscaled from 151 villages to 446 village clusters
    - Prepared 36 district Agriculture Contingency Plans (DACPs) totaling to 650 districts
    - Developed drought proofing action plans for 24 districts (16 in Karnataka, 4 each in Rajasthan and Andhra Pradesh)
- As much as 10.5 Cr. mobile-based agro advisories issued benefitting farmers and another clientele
    - Established 1949 units of agro-enterprises benefitting 3790 rural youth under ARYA program.
- Five New KVKs established in Sangli, Nagpur districts of Maharashtra; Kondagaon district of Chhattisgarh; and Hardoi and Moradabad districts of Uttar Pradesh
    - Established District Agro-Met Units (DAMUs) in 199 KVKs.
- 3 MoUs signed: AVRDC (World Vegetable Centre), Taiwan; International Centre for Bamboo Research, China, and Heinrich-Heine University, Germany for International Cooperation with international agencies and Universities.
- ICAR represented India and participated in the CGIAR System Council Meeting held in China.

Deemed Universities

Sl. No.	Institute	Abbre-viation	Located at	Establishment Year
1.	Indian Agricultural Research Institute	IARI	New Delhi	1905 (Status of Deemed University in 1958)
2.	National Dairy Research Institute	NDRI	Karnal, Haryana	1923 (Status of Deemed University in 1989)
3.	Indian Veterinary Research Institute	IVRI	Izatnagar, Uttar Pradesh	1889 (Status of Deemed University in 1983)
4.	Central Institute on Fisheries Education	CIFE	Mumbai, Maharashtra	1961 (Status of Deemed University in 1989)

Research Institutes

Sl. No.	Institute	Abbre-viation	Located at	Establishment Year
1.	Indian Council of Agricultural Research	ICAR	New Delhi	1929
2.	National Rice Research Institute	NRRI	Cuttack, Odisha	1946
3.	Indian Institute of Horticultural Research	IIHR	Bangalore, Karnataka	1967
4.	Indian Institute of Vegetable Research	IIVR	Varanasi, Uttar Pradesh	1971
5.	Central Island Agricultural Research Institute	CIARI	Port Blair	1978
6.	Central Arid Zone Research Institute	CAZRI	Jodhpur, Rajasthan	1952
7.	Central Avian Research Institute	CARI	Izatnagar, Uttar Pradesh	1979

Sl. No.	Institute	Abbre-viation	Located at	Establishment Year
8.	Central Inland Fisheries Research Institute	CIFRI	Barrackpore, West Bengal	1947 (Full Fleshed Institute in 1959)
9.	Central Institute of Brackishwater Aquaculture	CIBA	Chennai, Tamil Nadu	1987
10.	Central Institute for Research on Buffaloes	CIRB	Hisar Haryana	1985
11.	Central Institute for Research on Goat	CIRG	Makhdoom, Mathura, Uttar Pradesh	1979
12.	Central Institute of Agricultural Engineering	CIAE	Bhopal, Madhya Pradesh	1976
13.	Central Institute for Arid Horticulture	CIAH	Bikaner, Rajasthan	1994
14.	Central Institute of Cotton Research	CICR	Nagpur, Maharashtra	1976
15.	Central Institute of Fisheries Technology	CIFT	Cochin, Kerala	1957
16.	Central Institute of Freshwater Aquaculture	CIFA	Bhubaneshwar, Odisha	1987
17.	Central Institute of Research on Cotton Technology	CIRCOT	Mumbai, Maharashtra	1924
18.	Central Institute of Sub-Tropical Horticulture	CISH	Lucknow, Uttar Pradesh	1984 Renamed as CISH in 1995
19.	Central Institute of Temperate Horticulture	CITH	Srinagar	1994
20.	Central Institute on Post Harvest Engineering and Technology	CIPHET	Ludhiana, Punjab	1989*
21.	Central Marine Fisheries Research Institute	CMFRI	Kochi, Kerala	1947
22.	Central Plantation Crop Research Institute	CPCRI	Kasaragod, Kerala	1970

Sl. No.	Institute	Abbre-viation	Located at	Establishment Year
23.	Central Potato Research Institute	CPRI	Shimla, Himachal Pradesh	1949*
24.	Central Sheep and Wool Research Institute	CSWRI	Avikanagar, Rajasthan	1962
25.	Indian Institute of Soil and Water Conservation	IISWC	Dehradun, Uttrakhand	1974
26.	Central Soil Salinity Research Institute	CSSRI	Karnal, Haryana	1969
27.	Central Tobacco Research Institute	CTRI	Rajahmundry, Andhra Pradesh	1947
28.	Central Tuber Crops Research Institute	CTCRI	Trivandrum, Kerala	1963
29.	ICAR Research Complex for Eastern Region	ICAR-RCER	Patna, Bihar	2001
30.	ICAR Research Complex for North Eastern Region	ICAR-RC REH	Barpani, Meghalaya	1975
31.	Central Coastal Agricultural Research Institute	CCARI	Goa	1976
32.	Indian Agricultural Statistics Research Institute	IASRI	New Delhi	1930*
33.	Indian Grassland and Fodder Research Institute	IGFRI	Jhansi, Uttar Pradesh	1962
34.	Indian Institute of Agricultural Biotechnology	IIAB	Ranchi, Jharkhand	2014
35.	Indian Institute of Natural Resins and Gums	IINRG	Ranchi, Jharkhand	1924
36.	Indian Institute of Pulses Research	IIPR	Kanpur, Uttar Pradesh	1966*
37.	Indian Institute of Soil Science	IISS	Bhopal, Madhya Pradesh	1988
38.	Indian Institute of Spices Research	IISR	Calicut, Kerala	1995
39.	Indian Institute of Sugarcane Research	IISS	Lucknow, Uttar Pradesh	1952

Sl. No.	Institute	Abbre-viation	Located at	Establishment Year
40.	National Academy of Agricultural Research and Management	NAARM	Hyderabad, Telangana	1976
41.	National Institute of Biotic Stress Management	NIBSM	Raipur, Chhattisgarh	2012
42.	National Institute of Abiotic Stress Management	NIASM	Malegaon, Maharashtra	2009
43.	National Institute of Animal Nutrition and Physiology	NIANP	Bengaluru, Karnataka	1995
44.	Central Research Institute of Dryland Agriculture	CRIDA	Hyderabad, Telangana	1985
45.	Central Research Institute for Jute and Allied Fibers	CRIJAF	Barrackpore, West Bengal	1953
46.	National Institute of Natural Fiber Engineering and Technology	NINFET	Kolkata, West Bengal	1939*
47.	National Institute of Veterinary Epidemiology and Disease Informatics	NIVEDI	Bengaluru, Karnataka	2000
48.	Sugarcane Breeding Institute	SBI	Coimbatore, Tamil Nadu	1912
49.	ICAR- Vivekanand Parvatiya Krishi Anusandhan Sansthan	ICAR-VPKAS	Almora, Uttarakhand	1924 (Renamed in 1974)
50.	Central Institute for Research on Cattle	CIRC	Meerut, Uttar Pradesh	1987
51.	National Institute of High Security Animal Diseases	NIHSAD	Bhopal, Madhya Pradesh	2014
52.	Indian Institute of Maize Research	IIMR	New Delhi	1957*
53.	Central Agroforestry Research Institute	CAFRI	Jhansi, Uttar Pradesh	1988
54.	National Institute of Agricultural Economics and Policy Research	NIAEPR	New Delhi	1991
55.	Indian Institute of Wheat and Barley Research	IIWBR	Karnal, Haryana	1965*

Sl. No.	Institute	Abbre-viation	Located at	Establishment Year
56.	Indian Institute of Farming System Research	IIFSR	Modipuram, Uttar Pradesh	2009
57.	Indian Institute of Millet Research	IIMR	Hyderabad, Telangana	1958*
58.	Indian Institute of Oilseed Research	IIOR	Hyderabad, Telangana	1967
59.	Indian Institute of Oil palm Research	IIOPR	Pedavegi, West Godavari	1995
60.	Indian Institute of Water Management	IIWM	Bhubaneshwar, Odisha	1988
61.	Indian Institute of Rice Research	IIRR	Hyderabad, Telangana	1965
62.	Central Institute for Women In Agriculture	CIWA	Bhubaneshwar, Odisha	1996
63.	Central Citrus Research Institute	CCRI	Nagpur, Maharashtra	1985
64.	National Institute for Plant Biotechnology	NIPB	New Delhi	1985
65.	Indian Institute of Seed Research	IISR	Mau, Uttar Pradesh	2004

ICAR-National Research Centers

Sl. No.	Institute	Abbre-viation	Located at	Establishment Year
1.	National Research Centre for Banana	NRCB	Trichy, Tamil Nadu	1993
2.	National Research Centre for Grape	NRCG	Pune, Maharashtra	1997
3.	National Research Centre for Litchi	NRCL	Muzaffarpur, Bihar	2001
4.	National Research Centre for Pomegranate	NRCP	Solapur, Maharashtra	2005
5.	National Research Centre on Camel	NRCC	Bikaner, Rajasthan	1984
6.	National Research Centre on Equines	NRCE	Hisar, Haryana	1985

Sl. No.	Institute	Abbre-viation	Located at	Establishment Year
7.	National Research Centre on Meat	NRCM	Hyderabad	1986 in Izatnagar Shifted in Hyderabad 1999
8.	National Research Centre on Mithun	NRCM	Medziphema, Nagaland	1988
9.	National Research Centre on Orchids	NRCO	Pakyong, Sikkim	1996
10.	National Research Centre on Pig	NRCP	Guwahati, Assam	1990
11.	National Research Centre on Seed Spices	NRCS	Ajmer, Rajasthan	2000
12.	National Research Centre on Yak	NRCY	West Kemang	1989
13.	National Centre for Integrated Pest Management	NCIPM	New Delhi	1988
14.	Mahatma Gandhi Integrated Farming Research Institute	MGIFRI	Motihari, Bihar	2015

National Bureaues

Sl. No.	Institute	Abbre-viation	Located at	Establishment Year
1.	National Bureaus of Plant Genetics Resources	MBPGR	New Delhi	1977*
2.	National Bureaus of Agriculturally Important Micro-Organism	NBAIM	Mau, Uttar Pradesh	2001
3.	National Bureaus of Agricultural Insect Resources	NBAIR	Bengaluru, Karnataka	1957
4.	National Bureaus of Soil Survey and Land Use Planning	NBSS-LUP	Nagpur, Maharashtra	1976

Sl. No.	Institute	Abbre-viation	Located at	Establishment Year
5.	National Bureaus of Animal Genetic Resources	NBAGR	Karnal, Haryana	1984 In Bengaluru. Shifted to Karnal In 1985
6.	National Bureaus of Fish Genetic Resources	NBFGR	Lucknow, Uttar Pradesh	1983

Directorates/Project Directorates

Sl. No.	Institute	Abbre-viation	Located at
1.	Directorate of Groundnut Research	DGR	Junagarh
2.	Directorate of Soybean Research	DSR	Indore, Madhya Pradesh
3.	Directorate of Rapeseed and Mustard Research	DRMR	Bharatpur, Rajasthan
4.	Directorate of Mushroom Research	DMR	Solan, Himachal Pradesh
5.	Directorate of Onion and Garlic Research	DOGR	Pune, Maharashtra
6.	Directorate of Cashew Research	DCR	Puttur
7.	Directorate of Medicinal and Aromatic Plant Research	DMAPR	Anand, Gujarat
8.	Directorate of Floricultural Research	DFR	Pune, Maharashtra
9.	Directorate of Weed Research	DWR	Jabalpur, Madhya Pradesh
10.	Directorate of Poultry Research	DPR	Hyderabad
11.	Directorate of Knowledge Management in Agriculture	DKMA	New Delhi
12.	Directorate of Cold-Water Fisheries Research	DCWFR	Bhimtal, Nainital, Uttarakhand
13.	Project Directorate on Foot and Mouth Disease	PDFMD	Mukteshwar

State Agricultural Universities (ICAR Accreditation, 2019-20)

Sl.No.	Name of University
1.	Acharya N.G. Ranga Agricultural University, Guntur
2.	Acharya Narendra Deva University of Agriculture and Technology, Ayodhya
3.	Agriculture University Jodhpur, Jodhpur
4.	Agriculture University Kota, Kota
5.	Anand Agricultural University, Anand
6.	Assam Agricultural University, Jorhat
7.	Banda University of Agriculture and Technology, Banda
8.	Bidhan Chandra Krishi Viswavidyalaya, Mohanpur
9.	Bihar Agricultural University, Bhagalpur
1.0	Bihar Animal Sciences University, Patna
11.	Birsa Agricultural University, Ranchi
12.	Ch. Sarwan Kumar Krishi Vishvavidyalaya, Palampur
13.	Chander Shekhar Azad University of Agriculture and Technology, Kanpur
14.	Chaudhary Charan Singh Haryana Agricultural University, Hisar
15.	Chhattisgarh Kamdhenu Vishvavidyalaya, Durg
16.	Dr. Balasaheb Sawant Konkan Krishi Vidyapeeth, Dapoli
17.	Dr. Panjabrao Deshmukh Krishi Vidyapeeth, Akola
18.	Dr. Y.S.R. Horticultural Univeristy Venkataramannagudem,West Godavari
19.	Dr. Yaswant Singh Parmar University of Horticulture and Forestry, Nauni-Solan
20.	G B Pant University of Agriculture and Technology, Pantnagar
21.	Guru Angad Dev Veterinary and Animal Sciences University, Ludhiana
22.	Indira Gandhi Krishi Vishwavidyalaya, Raipur
23.	Jawahralal Nehru Krishi Vishwa Vidyalaya, Jabalpur
24.	Junagadh Agricultural University, Junagadh
25.	Kamdhenu University, Gandhinagar
26.	Karnataka Veterniary, Animal and Fisheries Sciences University, Bidar
27.	Kerala Agricultural University, Thrissur
28.	Kerala University of Fisheries and Ocean Studies, Panangad
29.	Kerala Veterinary and Animal Sciences University, Wayanad
30.	Lala Lajpat Rai University of Veterinary and Animal Sciences, Hisar
31.	Maharana Partap Horticultural University, Karnal
32.	Maharana Pratap University of Agriculture and Technology, Udaipur

Sl.No.	Name of University
33.	Maharashtra Animal and Fishery Sciences University, Nagpur
34.	Mahatma Phule Krishi Vidyapeeth, Rahuri
35.	Nanaji Deshmukh Veterinary Science University, Jabalpur
36.	Navsari Agricultural University, Navsari
37.	Orissa University of Agricultural and Technology, Bhubaneswar
38.	P.V. Narsimha Rao Telangana Veterinary University, Hyderabad
39.	Professor Jayashankar Telangana State Agricultural University, Hyderabad
40.	Punjab Agricultural University, Ludhiana
41.	Rajasthan University of Veterinary & Animal Sciences, Bikaner
42.	Rajmata Vijayaraje Scindia Krishi Vishwa Vidyalaya, Gwalior
43.	Sardar Krushinagar Dantiwada Agricultural University, Sardarkrushinagar
44.	Sardar Vallabhbhai Patel University of Agriculture and Technology, Meerut
45.	Sher-e-Kashmir University of Agricultural Sciences and Technology of Jammu, Jammu
46.	Sher-e-Kashmir University of Agricultural Sciences and Technology of Kashmir, Srinagar
47.	Sri Karan Narendra Agriculture University, Jobner
48.	Sri Konda Laxman Telangana State Horticultural University, Hyderabad
49.	Sri Venkateswara Veterinary University, Tirupati
50.	Swami Keshwanand Rajasthan Agricultural University, Bikaner
51.	Tamil Nadu Agricultural University, Coimbatore
52.	Tamil Nadu Dr. J. Jayalalithaa Fisheries University, Nagapattinam
53.	Tamil Nadu Veterinary and Animal Sciences University, Chennai
54.	U.P. Pt. Deen Dayal Upadhyaya Pashu Chikitsa Vigyan Vishwa Vidhyalaya Evem Go Anusandhan Sansthan, Mathura
55.	University of Agricultural and Horticultural Sciences, Shivamogga
56.	University of Agricultural Sciences, Bengaluru
57.	University of Agricultural Sciences, Dharwad
58.	University of Agricultural Sciences Raichur
59.	University of Horticultural Science, Bagalkot
60.	Uttar Banga Krishi Viswavidyalaya, Cochlear
61.	Uttarakhand University of Horticulture & Forestry, Bharsar
62.	Vasantrao Naik Marathawada Krishi Vidyapeeth, Parbhani
63.	West Bengal University of Animal and Fishery Sciences, Kolkata

Chapter 2

Agriculture Current Affairs – 2020-21

- ☆ Total food grain production of the country during 2018-19 – **285.20 million tons**
- ☆ Total food grain production of the country during 2019-20 (estimated) – **291.95 million tons**
- ☆ Total food grain production of the country during 2020-21 (targeted) – **293.30 million tons**
- ☆ Total production of Horticultural crops in India during 2018-19 – **310.74 million tons**
- ☆ Total production of Horticultural crops in India during 2019-20 (estimated) – **313.35 million tons**
- ☆ Total Geographical Area of the country- **329 Mha**
- ☆ Net sown area of the country- **140.82 Mha (42.57 per cent of total area)**
- ☆ Gross sown area of the country- **198.36 Mha**
- ☆ Barren land area in the country- **41 Mha**
- ☆ Degraded land area in the country- **63.85 Mha**
- ☆ Waste land area in the country- **178 Mha**

- ☆ Total forest cover in India is- **80.73 million ha (24.56 per cent of geographical area)**
- ☆ State with highest forest area in the country is- **Madhya Pradesh, followed by Arunachal Pradesh**
- ☆ For maintaining ecological balance percentage of the land area should be under forest is- **33**
- ☆ Irrigated area in India- **64.7 Mha (2019)**
- ☆ Total Micro irrigation area in India is – **11.58 Mha (Drip -5.75 Mha)**
- ☆ State with highest Micro irrigation area is- **Rajasthan> Andhra Pradesh> Maharashtra**
- ☆ State with highest area under Drip irrigation is- **Andhra Pradesh> MH> Gujarat**
- ☆ State with highest area under Sprinkler irrigation is- **Karnataka> Gujarat> Haryana>**
- ☆ Cropping Intensity of India is - **136 per cent**
- ☆ Government of India declared as National year of Millets - **2018**
- ☆ India is the largest producer of Millets in world, harvesting about 11 million tones and accounts for 40 per cent of the world's output.
- ☆ Contribution of Livestock to National GDP is- **5.1 per cent**
- ☆ Contribution of Livestock to Total Agricultural GDP- **27 per cent**
- ☆ The United Nations General Assembly has declared 2019 as the- **International Year of Indigenous Languages.**
- ☆ The United Nations General Assembly has declared 2020 as the- **International Year of Plant Health**
- ☆ The UN General Assembly adopted resolution designating 2021 as - **"International Year of Fruits and Vegetables".**
- ☆ The first state in the country to flagged off India's First fruit train was- **Andhra Pradesh**
- ☆ A fruit train was flagged from Tadipatri railway station in A.P. carrying a load of 980 metric tons of locally grown bananas to Jawaharlal Nehru Port in Mumbai from where the consignment will be exported to Iran (Bananas exported under brand name **Happy Bananas**)
- ☆ A joint initiative of the Indian Meteorological Department (IMD) and Indian Council for Agricultural Research (ICAR) to bring high resolution 'Vernacular Agro advisory App' based on weather forecast to farmers is- **MEGHDOOT App (Aug, 2019)**

- ☆ National Research Centre for Chikoo/Sapota to be set up in- **Thane, Maharashtra**
- ☆ Power house of Nutrients/Nutri-Cereals/Climate resilient crops are – **Millets**
- ☆ **Total number of KVKs at present is- 716 (as of April, 2020)**
- ☆ Effective functioning and monitoring of KVKs, KVKs are divided into- **11 ATARI Zones**
- ☆ *Six Month Cross border Agri-tech Incubation programme launched jointly by **Pusa Krishi** Incubator, ICAR and embassy of Indo-Brazil is- MAITRI (Dec 9, 2019) (Imp)
- ☆ Percentage of Farm mechanization utilization in India is about- **40 per cent**
- ☆ Total Food grain production in the country during 2018-19 was- **284.95 MT** (*Kharif* production- 141.71 MT; *Rabi* Production- 143.24 MT)
- ☆ India's share in global total food grain production is - **25 per cent**
- ☆ Total Pulse production for the year 2018-19 is- **23.4 MT**
- ☆ Among Pulses the crop with highest production and largest area under cultivation is- **Bengal gram/Chick pea/Chana/gram (Second is Arhar/ Pigeon pea)**
- ☆ Among Oil seeds the crop with highest production and largest area under cultivation is – **Soya bean (Second is Groundnut)**
- ☆ International Rice Research Institute (IRRI) South Asia Regional Centre established at- **Varanasi, U.P. on Dec, 2018** - First Director **- Dr. Arvind Kumar**
- ☆ Rice Knowledge bank launched (an agriculture web portal) launched on- 28 Feb, 2019 at **Assam. Funded by World Bank.**
- ☆ Khadi and Village Industries Commission (KVIC) has opened its first Silk processing plant at- **Surendranagar, Gujarat**.
- ☆ MSP was first announced for Wheat crop in the year- **1966-67**
- ☆ The state that produces highest food grains is- **Uttar Pradesh**
- ☆ The State that produces highest Pulses is- **Madhya Pradesh**
- ☆ The state that ranks first in Oilseed production is- **Rajasthan**
- ☆ The First Ever World Pulse Celebrated on-**10 Feb,2019**
- ☆ Total number of regulated markets in India- **7320**

- Total number of APMCs (Agricultural Produce Market Committees)- **585**
- Total number of GRAMs (Gramin Retail Agriculture Markets)-**22,000**
- Model Contract farming Act bill passed on-**May, 2018**
- The first state to enact law on Contract Farming in India is- **Tamil Nadu**.
- India's First Agromet Forecast Centre Inaugurated at- **UAS, Dharwad, Karnataka.**
- Main objective is to give accurate reports on weather which would enable the farmers to protect the crops and get a good yield.
- First state in India to use Automated Weather Stations (AWS) to provide accurate report on Weather to farmers is- **Maharashtra**
- Ministry of Tribal Affairs launched Minimum Support Price for Minor Forest produces and Value Addition as component of **Van Dhan Scheme.**
- India ranks in terms of Cultivation of GM Crops/Biotech crops- **5th place**
- In January 2020 the states that faced severe attack by Locusts is- **Gujarat and Rajasthan**
- The third edition of Global Potato conclave was held at- Gandhinagar, Gujarat on **28-31 Jan, 2020.**
- Headquarters of International Solar Alliance- **Gurugram, Haryana**
- National Nutrition Mission launched on- **8 March, 2018**
- Total number of Operational Holdings in India – **146.45 million**
- Average land holding size is- **1.08 hectare** (The average size of operational holdings is highest in Nagaland (5 hectares) and lowest in Kerala (0.18 hectares).
- Gross Cropped Area in India - **198.36 million ha**
- Net Sown Area in India- **140.82 Mha (48 per cent of total area)**
- Percentage of population depending on agriculture for their livelihood is- **54 per cent**
- Percentage of workforce depend on agriculture is- **50 per cent**
- Gross Irrigated area in India is- **96.46 Mha** (29 per cent of total cultivated area)
- Net Irrigated area in India is- **68.38 Mha** (20 per cent of total irrigated area)
- Maximum number of operational (landholdings) holdings are in – **U.P.** (23.82 million hectare)

- ☆ Minimum number of operational (landholdings) holdings are in–**Sikkim** (72,000 hectare)
- ☆ Percentage of small and marginal landholdings (below 2 hectares) constituted is – **86.21 per cent**
- ☆ Percentage of rural women depend on agriculture for their livelihood- **84 per cent**
- ☆ Percentage of women landholders (Cultivators) in agriculture is – **13.87 per cent**
- ☆ Target year fixed by Govt. of India for doubling of farmers income is- **2022** (75th Independence of India) Chairman of the Committee **Ashok Dalwai**
- ☆ All agricultural programs and policies shift their focus from- Production oriented to Income oriented approach
- ☆ National Productivity Council (NPC) Soil Health Card Recommendations has led to a- decrement of **8-10 per cent in use of chemical fertilizers** as well as increased productivity by **5-6 per cent.**
- ☆ Asia's First Rice Technology Park established at- **Karnataka**
- ☆ Total Cold storage Capacity of our country is- **34 MT (Million tons)**
- ☆ Bt cotton area in India is – **12.66 Mha** (96 per cent of cotton area)
- ☆ Bt cotton area is Highest in **Maharashtra followed by Gujarat.**
- ☆ Bt cotton is the only GM/Transgenic crop cultivated in India)
- ☆ First livestock census (once 5 years) conducted in India in the year- **1919-20**
- ☆ India's largest states with highest livestock population are- **Uttar Pradesh > Rajasthan**
- ☆ Cotton: Medium Staple length (mm) of 24.5-25.5 and micronaire value of 4.3-5.1; Long Staple length (mm) of 29.5-30.5 and micronaire value of 3.5-4.3
- ☆ Sugarcane prices are based on a basic recovery rate of **10 per cent**
- ☆ Total area under organic certification in India is- **1.5 million ha**
- ☆ Total Organic production in India- **1.35 Million tons**
- ☆ India's First organic state- Sikkim (Declared on Jan 18, 2016); Second organic state- **Uttarakhand**
- ☆ The country with largest area under organic farming – **Australia (27.15 Mha)**

- ☆ The country with highest number of organic producers in the world – **India** (More than 30 per cent of world's organic producers are in India)
- ☆ India's rank in terms of organically cultivated area is- **9th; First rank in terms of organic producers.**
- ☆ World's first wholly Organic country is- **Bhutan**
- ☆ Largest exported organic product in India- **Oilseeds (50 per cent)> Processed food products>Cereals and millets> Tea>Pulses>Spices**
- ☆ National Organic Farming Research Institute (NOFRI) in- **Sikkim** National Centre for Organic Farming - **Ghaziabad, U.P.**
- ☆ India's First Organic farming University going to be set up in- **Vadodara, Gujarat**
- ☆ Indian Institute of Farming System Research - **Modipuram, U.P.**
- ☆ Highest number of KVKs are in- **Uttar Pradesh**
- ☆ Technology Assessment and Refinement done through- **On Farm Trails (OFTs)**
- ☆ Krishi Kalyan Abhiyan (KKA): Launched on **1st June, 2018** It was launched to aid assist and advice farmers on how to improve their farming techniques and raise farm incomes. It was coordinated and implemented by KVKs in selected Aspirational districts (115 Districts). KKA will be implemented in 25 Villages of the Aspirational districts with more than 1000 population.
- ☆ mKisan Portal: Launched on 2013 provides Web-based mobile agro advisories (Weather, Market, Outbreak of pest and diseases and their control) to farmers with technological backstopping from KVKS and Research Institutes of SAUs)
- ☆ Government announces minimum support prices (MSPs) for 22 mandated crops and fair and remunerative price (FRP) for sugarcane. Total= 23 (22 mandated crops+ FRP for Sugar cane). The mandated crops are 14 crops of the *Kharif* season *viz.* paddy, jowar, bajra, maize, ragi, arhar, moong, urad, groundnut-in-shell, soyabean, sunflower, sesamum, Niger seed and cotton; 6 *Rabi* crops *viz.* wheat, barley, gram, masur(lentil), rapeseed/ mustard and safflower and two other commercial crops *viz.* jute and copra.
- ☆ Government of India for Water conservation and Water security with an emphasis on 1592 water stressed blocks in 256 districts is- Jal Shakti Abhiyan (JSA)- **1 July, 2019.**
- ☆ Number of Organic farming models developed by ICAR is- **45**

- ☆ Number of Integrated farming Systems validated by ICAR is- **51**
- ☆ Total Number of Biofortified varieties developed by ICAR is- **18** (Biofortified varieties are rich in Iron, Zinc, Protein, glucosinolates and Kunitz Trypsin Inhibitor (KTI) free.
- ☆ National Innovation Centre will be established by ICAR in- **New Delhi, 2020-21**. Innovations will be scientifically validated and farmers will be allowed to pursue research work.
- ☆ Currently ICAR is researching Zero Budget Natural Farming Methods (ZBNF) on- **Wheat and Basmati Rice**
- ☆ ICAR launched Deen Dayal Upadhyay Unnat Krishi Siksha Yojana (PDDUUKSY) - to train farmers in Organic farming, natural farming and cow-based economy.
- ☆ ICAR launched the Agricultural Education Portal for Centralised uniform solution called- **EKTA portal (Ekikrit Krishi Shiksha Tanki Ayaam)**
- ☆ National Phenomics Research Center established at- **New Delhi**
- ☆ Genetically Modified Cotton variety developed by Punjab Agricultural University (PAU), Ludhiana is– **PAU Bt 1 and F1861**
- ☆ Genetically Modified Cotton variety developed by Rajasthan Agriculture University is- **RS 2013**
- ☆ International Horticulture Innovation and Training Centre is located in - **Jaipur, Rajasthan**
- ☆ IARI has developed '**Pusa-** Soil Testing and Fertilizer Recommendation Meter' for facilitating the reach of soil testing at the farmers' doorstep.
- ☆ Gaint Leaf of *Macaranga nicobarica* holds potential as an alternative to plastics food wrappers/plates.
- ☆ Godetia: **A source of natural dyes**
- ☆ RAWE programme formulated by - M.S. Randwa committee for stipend provided for students is Rs.3,000
- ☆ Climate Smart Varieties developed by NRRI, Cuttack - **CR Dhan 801 and CR Dhan 802** (which are tolerant to both submergence and drought and few biotic stresses)
- ☆ Two superior chickpea varieties developed for the first time by using Genomics Assisted Breeding is - "**Pusa Chickpea-10216" and "Super Annigeri 1.** (Note: **Pusa Chickpea 10216** developed by ICAR- IARI in collaboration with the ICRISAT).
- ☆ **Super Annigeri 1**" has been developed by UAS, Raichur (Karnataka) and in collaboration with ICRISAT.

- ☆ New Blockbuster wheat variety developed by IARI with high yield and protein content is- **HD 3226 or Pusa Yashasvi**
- ☆ India's first high oleic acid groundnut varieties developed by ICAR-DGR, Junagadh is– **Girnar 4(ICGV 15083) and Girnar 5 (ICGV 15090)**
- ☆ Anti Oxidant rich Purple Potato Variety developed by CPRI, Shimla is- **Kufri Neelakanth**
- ☆ Latest Wheat Variety developed by ICAR for irrigated timely sown conditions of NE Plains is -**Karan Vandana (DBW-187)**
- ☆ Outstanding Forage Hybrid released in 2019 for revolutionizing forage sorghum production in India is- **CSH-24** (developed by GPBUAT and ICRISAT)
- ☆ Protein rich rice varieties - **CR Dhan 310 and CR Dhan 311**
- ☆ Pigeon pea -**PUSA Arhar 16** (determinate, extra early maturity, 120 days)
- ☆ Wonder variety of Sugarcane- **Co-0238** (First wonder cane variety Co-205)
- ☆ Bt cotton varieties developed by PAU is: **F1861, PAU Bt 1**(for cultivation in Punjab and Haryana)
- ☆ Bt cotton variety developed by Rajasthan Agricultural University is- **RS2013**
- ☆ Quality Protein Maize (QPM) -**PUSA Breakfast Cereal**
- ☆ First Hybrid variety of Tobacco in India is- **GTH-1**
- ☆ Low neuro toxin varieties of grass pea are -**Rattan, Nirmal, Prateek**
- ☆ Spineless variety of Safflower- **NARI-6**
- ☆ First ever short duration Moong (<60 days)- **Virat (IPM 205-7)**
- ☆ BLB of rice disease resistant variety-**PUSA 1592, Punjab Basmati-3**
- ☆ Blast disease resistant rice variety-**PUSA 109**
- ☆ Nematode resistant variety of papaya- **Pusa Majesty**
- ☆ Day neutral and spineless variety of Okra- **Pusa Sawani**
- ☆ Black Variety of Carrot- **Pusa Asita**
- ☆ World earliest maturity variety of cotton genotype was identified by CICR scientist (2017) - **Yugank**
- ☆ Early maturing Summer Mungbean- **Virat**
- ☆ **Pusa Arhar 16-** dwarfpigeon pea, Short duration (120 day redgram variety (Miracle red gram variety)

- **Co-0238 (Karan 4)** - Wonder variety of sugar cane, released by Sugarcane Breeding Institute, Regional Centre, Karnal.
- **Pusa rudhira** is an improved variety of-Carrot
- Antioxidant rich, red coloured Okra variety developed by IIVR, Varanasi- **Kashi lalima**
- An Arsenic resistant high yielding, thin aromatic rice variety is- **Muktoshri (W.B.)**
- New Potato Varieties for Table Purpose- **Kufri Neelkanth, Kufri Mohan, Kufri Ganga and Kufri Lima 6.**
- Farmer FIRST: Farmer's Farm, Innovations, Resources, Science and Technology (FIRST**), - July, 2015**
- National Agriculture Higher Education Project- Sep, 2017 Funded by: World Bank, Implemented by: **ICAR**
- Top States in terms of forest cover as Percentage of total geographical area is- **Mizoram (85.41 per cent)>Arunachal Pradesh (79.65) > Meghalaya (76.33 per cent)**
- Total forest cover in India is- **80.73 million ha** (24.56 per cent of geographical area)
- National Food Security Act/bill passed in the year- **July, 2013**
- As per Food Security Ordinance (2013) or Food security bill, food grain's planned to distribute among household is/are – Rice, wheat and coarse grains to distributed per households under Food Security Ordinance (2013) – Rs. 3:2:1 kg per month is – 5 kg, amount of food grain distributed per family living BPL is-35 kg.
- NITI Aayog act as 2015- **National Institution for Transforming**
- Total consumption of fertilizers in India in 2018-19 is- **27.29 million metric tons**
- NPK use ratio in India in 2018-19 is- **7.1:2.7:1** (average optimum use of NPK ratio is- **4:2:1)**
- Per hectare consumption of fertilizers in India during 2018-19 is-**137.4 kg/ha**
- Per ha consumption of fertilizers highest in- **Telangana (245.29 kg/ha) > Punjab (224.49 kg/ha)> Haryana (224.46 kg/ha).**
- Per ha consumption of fertilizers Lowest in- **Kerala (36.38 kg/ha) followed by Rajasthan (60.73 kg/ha).**

- ☆ Total production of NP fertilizers in India during 2018-19- **17.93 MMT** (N fert- 13.34, P fert- 4.59 MMT)
- ☆ Mostly Imported fertilizer- **Potash from Jordan and Tunesia**
- ☆ India is third in fertilizer production and second in consumption (China is first in fertilizer consumption)
- ☆ Fertilizer Control order- **1985**
- ☆ Neem Coated Urea (45 kg bag)- incorporated as regular fertilizer in **- 2010**
- ☆ Indigenous manufacturing of Neem coated Urea started in **-2014**
- ☆ Neem coated Urea is sold at- 5 **per cent above the MRP to recover the cost of coating**
- ☆ Neem coated urea save urea by-**20-30 per cent**
- ☆ The basic retail price of Urea is- **Rs.5360 per ton**
- ☆ The basic retail price of 45kg Neem Coated Urea bag is- **Rs.242**
- ☆ Nutrient Based Subsidy: scheme started in the Year- **2010-11**
- ☆ Where a fixed rate of subsidy is announced on nutrients. The NBS covers three macro nutrients (Nitrogen, Phosphorus and Potassium) and two micro nutrients (Zinc and Boron)
- ☆ The major pesticides applications are found in- Rice and Cotton crops (1st Paddy, 2nd Cotton)
- ☆ Andhra Pradesh, Maharashtra and Punjab are top three states contributing to 45 per cent of pesticide consumption in India.
- ☆ Monthly Radio Programme hosted by P.M. of India is- **Mann ki Baat** (started on Oct.3, 2014)
- ☆ Highest Gross Sown Area- **U.P.>M.P.>MH**
- ☆ Highest Net Sown Area- **MH and M.P**
- ☆ Barren and uncultivated land highest in- **Rajasthan**
- ☆ Largest state in terms of Geographical area- **Rajasthan**
- ☆ Government of India Formulated – **7 point strategy for Doubling of Farmers Income by 2022"**
- ☆ **Cluster bean-** becomes the first agricultural product to enter commodities market.
- ☆ Plant diagnostic app "**Plantix**" developed by- PEAT GmBH in collaboration with ICRISAT, CIMMYT and CABI

- ☆ In 2020 Budget Agriculture is presented under the theme of - **Asipirational India**.
- ☆ In 2020 Budget Government formulated point strategy to boost Agriculture **-16 point**
- ☆ Budget for Agriculture and allied sectors is- **2.83 Lakh cr** (Agril-1.42 lakh cr; RD-1.22 Lakh cr)
- ☆ Kisan Rail Scheme: 1 Feb, 2020, Launched by: **Indian Railways through PPP**. To build a Seamless National Cold Supply Chain for perishables (Milk, Meat, Fish) through refrigerated coaches in Express and Freight trains.
- ☆ Dhanya Lakshmi Scheme: 1 Feb, 2020- SHGs and Women SHGs will be promoted to become Dhanya Lakshmi (Seed banks) using Mudra Yojana.
- ☆ Pradhan Mantri Fasal Bima Yojana; Jan 14, 2016 under this scheme farmers will have to pay
- ☆ 2 per cent premium fixed by insurance firms for *Kharif* crops and 1.5 per cent for *Rabi* crops and 5 per cent for Commercial and Horticultural crops.
- ☆ It is formed by Merging National Agricultural Insurance Scheme (1999-2000); Weather based Crop Insurance Scheme (1985), Restructured Weather Based Crop Insurance Scheme (RWBCIS)- *Kharif* 2016- aims to mitigate the hardship of the insured farmers against the anticipated crop loss resulting from adverse weather conditions relating to rainfall, temperature, wind, humidity *etc.*
- ☆ FASAL - Forecasting Agricultural output using Space, Agro meteorological and Land based observations (FASAL) - **2010**

  Aim was to collect Monsoon data through remote sensing, economic data and monitoring of crops when growing.
- ☆ National Mission for Sustainable Agriculture (NMSA)- **2014**
- ☆ National Mission on Micro Irrigation (NMMI)- **2010**
- ☆ "India is a member country of **SAARC, G-20, G-77, WTO, IMF, World bank and BRICS, BIMSTEC**". India is not a member country of **ASEAN, G-7, OPEC.**
- ☆ National Remote Sensing Centre- **Hyderabad, Telangana**
- ☆ Borlaug Institute for Asia - **Ludhiana, Punjab**
- ☆ National Multi Commodity Exchange- **Ahmadabad, Gujarat**
- ☆ National Bee Board- **New Delhi.**

Chapter 3

Agricultural Revolutions and Important Days Related to Agriculture and Allied Sciences

Revolutions Related to Agriculture

Revolutions	*Associated Crop*	*Father*
Green Revolution	Over all crop Production	Norman Borlaug- World M.S. Swaminathan- India
White Revolution	Milk Production	Verghese Kurien
Black Revolution	Petroleum Production/ crude oil production	Hiralal Chaudhary and Arun Krishnan
Brown Revolution	Leather/Cocoa Production/ non conventional	-
Golden Fibre Revolution	Jute Production	
Golden Revolution	Fruits/Overall Horticulture development/Honey Production	Dr. K.L. Chadda
Grey Revolution	Fertilizer	--
Brown Revolution	Leather/Cocoa production	--
Yellow Revolution	Oil Seeds	Sam Pitroda

Revolutions	***Associated Crop***	***Father***
Round Revolution	Potato	
Red Revolution	Meat and Tomato Production	Dr. Vishal Tewari
Pink Revolution	Onion, Prawn	Dr. Durgesh Patel
Protein Revolution	Agriculture Higher production Technology driven Green Revolution	Sri Narendra Modi
Silver Fiber Revolution	Cotton	--
Silver Revolution	Egg/Poultry	Smt. Indira Gandhi
Evergreen/Rainbow revolution	Overall development of Agriculture	M.S. Swaminathan
Sweet Revolution	Honey Production	Sri Narendra Modi
Green Gold Revolution	Promotion and Trade of Bamboo	Sri Narendra Modi

Important Days in Agriculture and Allied Sciences

World Wetland Day/Ramsers Day	2 February
World Pulses Day	10 February
World Radio Day (for Social sciences)	13 February
Soil Health Card Day	19 February
World Wildlife Day	3 March
International Women's Day	8 March
Water Resource Day	11 March
World Consumer Day	15 March
World Forest Day	21 March
World Water Day	22 March
World Meteorological Day	23 March
World Health Day	7 April
World Earth Day	22 April
World intellectual Property Day	26 April
World Veterinary Day	Last Saturday of April
National Technology Day	11 May
World Honey Day	20 May
World Tea Day	21 May
World Anti- Tobacco Day	31 May

World Milk Day	1 June
World Food Safety Day	7 June
World Environment Day	5 June
National Fish Husbandry Day	10 July
World Population Day	11 July
ICAR Day	16 July
World Literacy Day	8 Sep
World Honey Day	August 19
World Ozone Day	16 Sep
World Bamboo Day	18 Sep
World Habitat Day/Vegetarian Day	1 Oct
World Coffee Day	1 Oct
National Women Literacy Day	2 Oct
World Animal Welfare Day	4 Oct
International Day of Girl Child	11 Oct
World Egg Day/UN International Day for Natural Disaster Reduction	13 Oct
Mahila Kisan Diwas (Women in Agriculture Day)/ International Day of Rural Women	15 Oct
World Food Day	16 Oct
UN Day	24 Oct
National Press Day	16 Nov
World Fish Day	21 Nov
National Milk Day	26 Nov
National Agriculture Education Day (Babu Rajendra Prasad birthday)	3 Dec
World Soil Day	5 Dec
Human Right's Day	10 Dec
National Energy conservation Day	14 Dec
Farmers Day (Kisan Diwas) by Ch.Charan Singh birthday Farmers awareness week (Dec 23-29)	23 Dec

Chapter 4

Agronomy: Important Terminologies

A Value: It is defined as available soil nutrient determined in terms of a standard fertilizer used. It is used for assessment of available P and S in soils, expressed as

$$A = \frac{B(1-y)}{y}$$

Where B is the amount of nutrient in the applied fertilizer plot and y is proportion of the nutrient in the plant derived from fertilizer. The "A value" can only be determined by tracer technique using a labelled fertilizer it is used in research, and not for farm advisory purposes.

Abrasion: Breakdown of clods, crusts, and plant material by the impact of particles moved by wind in saltation. The impacting particles may also abrade. Abrasion causes soil aggregates to break down progressively as wind erosion continues.

Abscissa: The distance from the vertical axis (Y-axis) measured parallel to the horizontal axis (X-axis) of a graph in Cartesian Co-ordinate system.

Absolute Density: Density expresses the numerical strength of a species in a plant community. It is a parameter used by the ecologists to measure the number of individuals per unit area.

$$= \frac{\text{Total number of individuals of a species in all quadrats}}{\text{Total number of quadrats studied} \times \text{Areas in m}^2 \text{ of a quadrat}}$$

Absolute Frequency: It refers to the degree of uniformity of occurrence or dispersion of a species of plants in an area and represents the number of sampling units in which that particular species occurred. It is expressed as percentage of the total number of samples/quadrats examined.

$$= \frac{\text{Number of quadrats in which a species occurred}}{\text{Total number of quadrats studied}} \times 100$$

Absolute Growth Rate: The rate of increase in size of a growing plant (or part of it) in a given time under specific conditions.

Absorption: Plants absorb water and also the substances dissolved in it including nitrogen and other mineral elements, largely through root hairs. The root hairs absorb water by osmosis. The more a plant needs water the more vigorously it is absorbed, provided the water supply remains ample. It is also possible for water to be extracted from the roots, as can happen in the case of highly saline soils and saline soil solution.

Abundance: It is the number of individuals of different species in a community per unit area based on the quadrats where only they occurred. It cannot provide the numerical strength of a species in an area since all quadrats are not taken into consideration. However, abundance is useful in determining the dominance of a species and can be calculated as follows:

$$= \frac{\text{Total number of individuals of a species in all (N) the quadrats}}{\text{Total number of quadrats (n) in which the species occurred}} = \frac{\sum Wi}{n}$$

Where, ΣWi is the sum of individuals of a species occurring in all the quadrats, *i.e.* in "N" number of quadrats, but here actually in "n" number of quadrats.

ACCase Inhibitors: Herbicides which inhibit acetyl coenzyme A carboxylase (ACCase), a multi-functional, bio-tinylated protein located in stroma of plastids and catalyses ATP-dependent condensation/carboxylation of acetyl COA and bicarbonate (HCO_3) to malonate in fatty acid biosynthesis pathway. They are aryl-oxy-phenoxy-propinates and cyclo-hexane-diones herbicides.

Accelerated Erosion: Erosion of soil resulting from disturbance of the natural landscape. It results largely from the consequences of human activity, such as tillage, grazing, and removal of vegetative cover.

Acclimatization: It is the process in which an individual organism adjusts to a change in its environment (such as a change in altitude, temperature, humidity, photoperiod, or pH), allowing it to maintain performance across a range of environmental conditions.

An adjustment in the morphology or physiology of an organism in response to a change in the environment. In other words, acclimatization refers to physiological and behavioural adjustment of a living organism to change in its immediate environment.

Acid Equivalent (a.e.): Acid equivalent is the amount of toxic material in terms of parent acid of a herbicide present in the commercial formulated product. The theoretical yield of parent acid forms an active ingredient.

Acid Rain: Rain water containing excessive concentration of acidic components, primarily NO_3^-, SO_4^-, H^+, having pH of 5-7.

Acid Soil: A soil with a pH of less than 7.0 (usually less than 6.6). An acid soil has a preponderance of hydrogen ions over hydroxyl ions.

Acid Sulphate Soil: Very acid soil (pH < 4) in which sulphuric acid is formed by oxidation of S bearing pyrite minerals. Found primarily in coastal, deltaic and estuarine areas of the humid tropics. Sometimes also called cat clays.

Acid Value: It is the amount of potassium hydroxide required in mg to neutralise the free acids present in one gram of the substance.

Acid: A substance which forms hydrogen ions (H^+) when dissolved in water, a proton donor.

Acidification: Acidification is a natural process that occurs continuously in soils. It is caused by the following factors: The breakdown of organic matter can cause acidification of the soil as amino acids are converted into acetic acid, hydrogen gas, dinitrogen gas, and carbon dioxide by the reaction:

- ☆ The movement of acidic water from rainfall through soils slowly leaches basic essential elements such as Ca, Mg, and K below the plant root zone and increases the concentration of exchangeable soil Al. Soluble Al_3^+ reacts with water to form this equation, which makes the soil acid.
- ☆ Soil erosion removes exchangeable cations adsorbed to clay particles.
- ☆ Hydrogen is released into the soil by plant root systems as a result of respiration and ion uptake processes during plant growth.
- ☆ Nitrogen fertilization speeds up the rate at which acidity develops, primarily through the acidity generated by nitrification.

Acropetal: Acropetal order is found in racemose inflorescence and it means that older flowers are at the base and the younger flowers are at the tip.

Active Ingredient (a.i.): It is the pure and concentrated form of toxic herbicide present in the formulated product, which kills weeds or brings about inhibition in their growth. It is expressed in percentage of the commercial product and its value is written always on the label and container of an herbicide. Thus, a commercial formulation, isoguard 75 WP, states that it contains 75 per cent a.i. and the formulation is wettable powder.

Active Uptake: Nutrient absorption mechanism for which energy is needed. Often carried out with a participation of a carrier, which transports the nutrients across cell membrane.

Acute Toxicity/Poisoning: If the symptoms of poisoning occur immediately after exposure, it is called acute toxicity or poisoning. It is believed that the pesticide has acted in no time.

Additivity or Additive Effect: It refers to the combined action of the component herbicides in mixture when the total effect of mixture is equal to the sum total effects of the component herbicides applied independently.

Adhesive: A material included in formulation to increase its sticking power or adhesiveness. Also known as a sticker.

Adiabatic Lapse Rate: The adiabatic lapse rate is the rate at which the temperature of an air parcel changes in response to the compression or expansion associated with elevation change, under the assumption that the process is adiabatic, *i.e.*, no heat exchange occurs between the given air parcel and its surroundings. The rate is constant for dry air, *i.e.* 10^0C per kilometre.

Adjuvant: An adjuvant is a substance when mixed in a formulation improves spreading, wetting, penetration and translocation, facilitates slow release, reduces drift and evaporation *etc.* An ingredient which when added to a formulation aids the action of the toxicant. Includes such materials as wetting agents, spreaders, adhesives, emulsifiers, dispersing agents and correctives.

Adsali sugarcane: Sugarcane which takes 18 months for harvesting, usually planted in June-July.

Adsorption: The process by which atoms, molecules, or ions are taken up from the soil solution or soil atmosphere and retained on the surfaces of solids by chemical or physical binding.

Adsorption: It is the holding up of herbicide molecules, ions onto the surface of soil colloids by physic-chemical attraction. This leads to less availability of herbicides in the soil solution and in turn less uptake by plants.

Adventitious Roots: Roots originating from any part or organ of the plants other than radicle are adventitious root, *e.g.* fibrous roots of the *Poaceae* plants.

Adventive Weeds: The exotic or introduced weeds, which need further human intervention for their continued existence in a particular region or country.

Aeoline: A sediment deposit carried by wind.

Aerobic Respiration: Enzymatic oxidation of organic substances into a simple component.

Aerobic Rice: Aerobic rice is a production system in which especially developed "aerobic rice" varieties are grown in well drained, non puddle unsaturated soil. Emits 80-85 per cent lesser methane gas into atmosphere and less cost of production due to reduction in land preparation and no transplanting.

Aerobic Soil: Well aerated soil, having adequate free oxygen. Most upland non flooded, adequately drained soil are aerobic.

Aerobic: Condition with oxygen gas as a part of the environment, living or active only in the presence of air or free oxygen.

Aflatoxins: These are naturally mycotoxins in grains/seeds/kernels that are produced by *Aspergillus flavus* and *Aspergillus parasiticus,* species of fungus. The food quality standards on aflatoxins are different on different countries and usually vary from 5- 30 PPB.

Aggregate Stability: The ability of a soil aggregate to resist various destructive forces, such as tillage, abrasion by wind or flowing water, or raindrop force.

Aggregation Soil: The cementing or binding together of primary soil particles (sand, silt, and clay) into a secondary unit, which unit contributes to the soil structure.

Aggressivity (A): Gives a simple measure of how much the relative yield increase in species A is greater than that for species B in an intercropping system. It can be expressed as

$$Aab = (Yab/(Yaa \times Zab)) - (Yba/(Ybb \times Zba))$$

Where Yaa = pure stand yield of species A, Ybb = pure stand yield of species B, Yab = mixture yield of species A with combination with B, Yba = mixture yield of species B in combination with species A, Zab = sown proportion of species A in mixture with species B, Zba = sown proportion of species B in mixture with species A.

Agricultural Drought: It occur where soil moisture and rainfall are inadequate during the growing season to support healthy crop growth to maturity and process extreme crop stress and wilt.

Agricultural soil/furrow-slice/plough-sole layer: It is the stratum/profile/layer of soil from surface to a depth of 15 cm ("6") up to which ploughing is usually done and where maximum rhizosphere activity of crops/plants takes place. The amount of soil is $2.24*10^6$ kg/ha.

Agricultural Statistics: Agricultural statistics is a science that deals with the methods for drawing valid conclusions and making decisions under uncertainty through generation of quality data related to agricultural research and planning.

Agrisilviculture: It is defined as growing of trees and agriculture crops together in same lands at the same time.

Agroclimatic Region: For resource development, the country has been broadly divided into fifteen agricultural regions based on agroclimatic features, particularly soil type, climate including temperature and rainfall and its variation and water resources availability.

Agroecology: A sub-discipline of geology on the scientific study of the origin, nature, composition, distribution and utilization of soils form a geological view point.

Agrology: The study of applied phase of soil science and soil management.

Agronomic Efficiency: Denotes the units of additional crop produced per unit of input added externally. *e.g.* kg grain/kg of nutrients added. Agronomic efficiency has a direct bearing on better utilization of applied nutrients, the profitability of their usage and minimisation of nutrient losses.

Agronomic Management Index (AMI): It is calculated as follows:

$$= \frac{\text{Per cent crop yield over control} - \text{Per cent control of weeds or other pest}}{\text{Per cent control of weeds or orther pests}}$$

Agronomic Rate: The rate at which fertilizers, organic wastes or other amendments can be added to soils for optimum plant growth.

Agronomy: Specialized branch of agriculture dealing with crop production and management. This is the application of scientific principles to the art of crop production.

Agrostology: Also known as graminology, is the scientific study of the grasses.

AHAS/ALS inhibitors: Herbicides which inhibit aceto-hydroxy acid synthase (AHAS)/aceto-lactate synthase (ALS) enzyme, which catalyzes the synthesis of three branched-chain amino acids, *e.g.* leucine, isoleucine and valine in two related or parallel reactions. They include sulfonylures, imidazolinones, tri-azolo-pyrimidine sulfonanilides and pyrimidinyl-thio-benzoates.

A-horizon: The surface horizons of a mineral soil having maximum biological activity or eluviation's (removal of materials dissolved or suspended in water) or both.

Air-Dry Weight: Weight of a substance after it has been allowed to dry to equilibrium with the atmosphere.

Akiochi Soil: These are flooded soils were hydrogen sulphide is formed due to sulphate reduction and anaerobic decomposition of organic matter.

Albedo: Albedo is the measure of the diffuse reflection of solar radiation out of the total solar radiation.

Alfisols: Soil with grey to brown surface horizons, medium to high supply of bases, and B horizons of illuvial clay accumulation.

Alkaloids: Bitter alkaline components, containing carbon, hydrogen, nitrogen and some oxygen. Example caffeine, morphine, nicotine and quinine.

Allelopathy: Production of a substance by one organism that inhibits one or more other organisms.

Alley Cropping: Alley Cropping is planting rows of trees at wide spacings with a companion crop grown in the alleyways between the rows.

Allochory: It is the mechanism of weed seed dispersal solely by several extraneous factors/agents like human, animal, wind, water, crop seed, FYM/compost/organic matter, *etc.*

Allophane: The most significant amorphous silicate mineral, poorly defined aluminium silicate ($Al_2O_3.2SiO_2. H_2O$), most prevalent in soils developed from volcanic ash and capacity to adsorb cations.

Alluvial Soil: A soil developed from recently deposited alluvium and exhibiting essentially no horizon development or modification of the recently deposited materials.

Alternative Hypothesis: The desired attitude of the researcher about the statistical hypothesis is termed as alternative hypothesis. It is usually denoted by H_1. It is a hypothesis about the population parameters the possibility of which not rejecting is of interest to the experimenter.

Aluminium Toxicity: It is caused by excess water soluble and exchangeable aluminium, usually on acid sulphate soils and strongly acid soils. Symptoms are white or yellow interveinal blotches on the leaves. The leaves dry out and die. Roots are short and scanty and the plants are stunted.

Aman: This variety is sown in the rainy season (July-August) and harvested in winter. This is mainly produced.

Amendment: Any material such as lime, gypsum, saw dust, sulphur or soil conditioner that is added to the soil to improve the physiochemical properties of the soil. A substance added to the soil to improve plant growth, such as lime.

Soil Amendment: A soil amendment is any material added to a soil to improve its physical properties, such as water retention, permeability, water infiltration, drainage, aeration and structure.

Ammonium Fixation: Absorption of ammonium ions by soils or minerals in such forms that they are neither water soluble nor readily exchangeable.

Amphipathic: It represents a property of a surfactant which indicates a surfactant possesses both L (polar) and H (polar) ends in a single molecule. The polar (hydrophilic) portion gives a surfactant its ionic properties, *i.e.* anionic, cationic and non-ionic depending upon the ionic species they release in solution.

Amphiphytes:

Plants that grow in two contrasting conditions such as on land and in water.

Anti-quality constituents of some plant species:

	Plant Species	***Anti-quality Constituents***
1	Sub-terranean clover, red clover, white clover	Castrogens
2	Sweet clover (*Melilotus* spp.)	Coumarins
3	Lucerne (Alfalfa), ladino clover, bur clover, strawberry clover, *trifolium*	Saponins

	Plant Species	Anti-quality Constituents
4	*Phalaris, Crotalaria, lupins, desmodium, lespedeza*	Alkaloids
5	*Sorghum* spp., *Trifoliumrepens, Lotus* spp., *poa aquatic, Zieria laevigata, Phyllanthus qastroemmi, Cyanodon plectostachyus, Petridium acqullinum*	Hydrocyanic acid (HCN)/ prussic acid/cyanogenetic glycosides
6	Oat, para grass	Nitrate
7	Napier grass, pearl millet	Oxalic acid
8	*Leucaena leucocephala*	Mimosine
9	*Trifolium* spp., white clover	Bloat-producing constituents-plant cytoplasmic proteins
10	Shrub and tree spp.	Tannins
11	Lentin	Raffinose

Anaerobic: Non-aerobic state or deficient in oxygen.

Analysis of Covariance (ANCOVA): This is an extension of analysis of variance and exploits the advantage that the analysis can provide by making use of auxiliary information. This is another way of controlling the experimental error by making use of auxiliary variable(s) in the model. The auxiliary variable(s) should be such that they are not affected by the application of treatments. The advantage will be significant only when the values of the auxiliary variable(s) vary over the experimental units. This is also useful when some observations are missing.

Analysis of Variance (ANOVA): It is a statistical technique to segregate the total variation in the sampled data into various groups with known causes of variation and another group with unknown causes of variation.

Animal Manure: Manure is an important agricultural by-product and may be a potentially valuable source of nutrients containing N, P, and K. Depending upon the type of manure and the characteristics of the animal feeding operation, it may also contain other essential nutrients such as calcium (Ca), sulphur (S), boron (B), magnesium (Mg), manganese (Mn), copper (Cu) and zinc (Zn). Although animal manure has historically been applied to agricultural land in areas.

Anion Exchange Capacity: The sum total of exchangeable anions that a soil can absorb, expressed as milliequivalents/100 g of soil.

Anion Exchange: Replacement of one anion by other on positively charged soil surfaces, *e.g.* exchange of $H_2PO_4^-$ with F^- and NO_3^- with Cl^-.

Annidation: The complimentary interaction between intercrops in the intercropping system is known as Annidation.

Anti-transpirants: Substances that retard or inhibit transpiration, *e.g.* PMA, kaolin *etc.*

Antogonistic Effect: *Antagonism* refers to the yield in response of two *nutrients* in which the combined *effect* is less than expected from the individual responses.

Apatite: A naturally occurring complex calcium phosphate which is the original source of most of the phosphate fertilizer.

Apogamy: Development of embryo from synergists or antipodal cells without fertilization.

Apoplast: Non living cells (xylem) forming a continuous system for movement of water in the plant body or non-living continuum in plants through intercellular space and cell walls.

Apparent Nutrient Recovery (ANR): Nutrient absorbed by a crop from applied fertilizer expressed as a percentage of nutrient applied. Usually measured without tracer technique and tends to over estimate fertilizer recovery.

$$= \frac{\text{Nitrogen uptake(kg/ha) in fertilized plot} - \text{Nutrient uptake (kg/ha) in control plot}}{\text{Fertilizer nutrient applied (kg/ha)}} \times 100$$

Application Efficiency (Ea): The percentage ratio of amount of water stored in the crop root zone to the amount of water delivered to the crop field.

Ea = (water stored in the root zone/water delivered to field) x 100.

APSIM Model: Agricultural production system simulations (APSIM) is a biophysical model developed to simulate biophysical processes in faming systems.

Aquaculture: Aquaculture, also known as aquafarming, is the farming of fish, crustaceans, molluscs, aquatic plants, algae, and other organisms. Aquaculture involves cultivating freshwater and saltwater populations under controlled conditions, and can be contrasted with commercial fishing, which is the harvesting of wild fish.

Argillation: It is the process by which the dispersed clay particles are migrated from upper to the lower horizons result a textural horizon.

Aridisols: Mineral soils that have an aridic moisture regime and ochrich epipedon and other pedogenic horizons but non oxic horizons (an ordered in the USDA soil taxonomy).

Aridity Index (AI): A measure of dryness of a region and its expressed as,

$$= \frac{\text{No. of rainy days} \times \text{Mean precipitation/day}}{\text{Mean temperaturee} + 10}$$

Arithmetic Mean: It is the ratio of sum of all values of a series to that of number of values in that series.

Arrowing: Flowering in sugarcane is referred as arrowing.

Assimilatory quotient: The assimilatory quotient (AQ), the ratio of net CO_2 consumption to net O_2 evolution.

Atmometer: An atmometer or evaporimeter is a scientific instrument used for measuring the rate of water evaporation from a wet surface to the atmosphere.

Atmosphile Elements: Which tend to remain in atmospheric gases.

Augmenting Crops: When different sub crops are sown for the purpose of supplement the yield of the main crop, then the sub crops are known as augmenting crops.

Aus: Rice sown in summer along with the pre-monsoonal showers and harvested in autumn. The quality of this rice is rough.

Auxanometer: An instrument which is used to measure growth of plants.

Auxiliary Bud: The axillary bud is an embryonic or organogenic shoot located in the axil of a leaf.

Avagadro's Number: The number of molecules in a gram molecular weight of any compound (6.023×10^{23}).

Available Nutrient: A portion of any element or compound in the soil that can be readily absorbed and assimilated by growing plants.

Available Water Capacity: Available water capacity is the traditional term used to express the amount of water held in the soil available for use by most plants. It is dependent on crop rooting depth and several soil characteristics.

Units of measure are expressed in various terms:

- Volume unit as inches of water per inch or per foot of soil depth
- Gravimetric percent by weight
- Percent on a volume basis

Available Water: Soil moisture between field capacity and permanent wilting point is referred to as readily available moisture. It is the moisture available for plant use.

Avalanching: The increase in rate of soil flow with distance downwind across an area being eroded by wind.

A-Value Technique: Radiochemical analysis of plants grown on soils which have been treated with fertilizer containing elements such as radioactive phosphorus; may be used to calculate the phosphorus supply of the original soil.

Azolla: A group of aquatic ferns capable of fixing high level of nitrogen from the atmosphere and are widely grown as a fertilizer crop in lowland rice cultivation system.

Azospirillum: A nitrogen fixing root and soil inhibiting bacterium in tropics.

Azotobacter: An aerobic, non-symbiotic nitrogen fixing bacterium.

B horizons: The sub soil layer in which certain leeched substances are deposited.

Bagasse: Bagasse is the dry pulpy fibrous residue that remains after sugarcane or sorghum stalks are crushed to extract their juice. It is used as a biofuel for the production of heat, energy, and electricity, and in the manufacture of pulp and building materials.

Balanced Fertilizer: A soil additive containing suitable proportion of each necessary mineral element to grow a crop or a plant.

Balanced Nutrition: A condition when supply of all plant nutrients is at an optimum level, there are no deficiencies, no excess and no antagonisms. An ideal condition is fertilizer management and crop nutrition.

Bale: A large wrapped or bound bundle of paper, hay, or cotton. 1 bale of cotton- 170 kg; 1 kg of Jute- 180 kg.

Band Application: An application to a continuous restricted band along a crop row rather than over the entire area, mostly done in case of total killer or non-selective herbicides. Saving in herbicide, water and cost of spraying are achieved.

Band Placement: Fertilizer placement in bands on one side or both side of the row above 5 cm below the seed and 5 cm to one side of the seed or plant grown, or around each plant.

Base (period): Period in days during which irrigation is supplied to a crop.

Base Exchange Capacity: The extent to which the soil can absorb exchangeable cations other than hydrogen and aluminium.

Basic Slag: A by-product of steel industry obtaining from phosphate iron ores. The containing about 6 -18 per cent phosphoric acid. Also contain 33.9 per cent Ca and 3 per cent S. Also called as Thomas slag.

Basin Listing: Tillage that forms lister furrows at regular intervals to create small basins to capture and store rain or applied water.

Basipetal: Basipetal order is found in cymose inflorescence and it means that older flowers are at the top and the younger flowers are at the base.

Basophiles: A plant which can grow well in alkali soils (pH 7.4 – 8.5).

Baule Unit: A unit of fertilizer or any other growth factor be taken as that amount necessary to produce a yield of 50 per cent of the maximum possible. Baule unit for N, 223 lb/acre; P_2O_5, 45 lb/acre; and K_2O, 76 lb/acre.

Bench Terrace: Bench terraces are a soil and water conservation measure used on sloping land with relatively deep soils to retain water and control erosion. They are normally constructed by cutting and filling to produce a series of level steps or benches. This allows water to infiltrate slowly into the soil.

Benefit: The present value o benefits divided by the present value of costs.

Beushining: Beushening, commonly practiced in submerged lowlands of eastern India, is a traditional cultural practice of cross-ploughing the dry-seeded (broadcast) standing crop of rice 25-35 days after seeding when 15-20 cm of rainwater gets impounded in rice fields, followed by laddering and seedling redistribution which facilitates stable rice yields under low levels of inputs and uncertain climatic conditions.

Bias: It is the difference between an estimator's expectation and the true value of parameter.

Binary Fertilizer: Contains two major nutrients *e.g.* Potassium Nitrate.

Binomial Distribution: It is a discrete probability distribution expressing the probability of one set of dichotomous alternatives that is success or failure. More precisely, the binomial distribution refers to a sequence of events which possess the following properties: (i) An experiment is performed under same conditions for a fixed number of trial assay, n, (ii) In each trial, there are only two possible outcomes of the experiment called success or failure, (iii) The probability of a success denoted by 'p' remains constant from trial to trial, (iv) The trials are independent *i.e.* the outcomes of any trial or sequence of trials do not affect the outcomes of subsequent trials.

Bioagent: A living organism employed to control a pest.

Bioassay (biological assay): Bio-assay combines "bio" means living system and "assay" means method/procedure. Thus, it represents a qualitative and quantitative method of determining the amount of chemicals present in soil employing a highly sensitive biotic system, plants or animals.

Biochemical Oxygen Demand (BOD): The amount of oxygen required by aerobic organisms to carry out oxidative metabolism in water containing organic matter, such as sewage. BOD is used as an indirect measure of the concentration of biologically degradable material present in organic wastes. Also known as Biological Oxygen Demand.

Biodegradation: Oxidative breakdown of synthetic oil and natural organic substances by microbial activity.

Biofertilizer: Preparation containing live or latent cells of efficient cells of nitrogen fixing, phosphate solubilizing or cellulolytic micro-organisms used for application to seed, soil or composting areas with the objective of increasing the number of such micro-organisms and accelerate those microbial processes which augment the availability of nutrient that can be easily assimilated by plants.

Biogas: A mixture of gases containing methane, carbon dioxide, hydrogen and traces of few other gases produced by anaerobic fermentation of easily decomposable cellulolytic materials in presence of methane producing bacteria.

Biomass: The total mass of living organisms in a given volume or mass of soil, or in a particular environment.

Biome: A large naturally occurring community of flora and fauna occupying a major habitat, *e.g.* forest or tundra.

Bioremediation: The use of biological agents to reclaim soil and water polluted by substances hazardous to the environment or human health.

Biostatic Effect: Reduced microbial decomposition of a pesticide in the presence of some chemical, called extender *e.g.* boron is biostatic to 2, 4-D.

Biosuper: A biological fertilizer consisting of rock phosphate, elemental sulphur and an inoculum of sulphur-oxidising *thiobacilli* was compared with superphosphate as a source of phosphorus.

Biotype: They are individuals of the same species of plants/weeds differing in physiological and biochemical response to an herbicide from rest of the individuals.

Biuret: A condensation upto molecule sand is toxic to many plants. Its content in the commercial fertilizer should not exceed 2 per cent.

Black Carbon: It is carbonaceous material produced by fire as in the case of slash and burn agriculture. Two forms of differentiating: (a) charcoal is composed of incomplete biomass combustion that represents the main portion of black carbon and (B) condensation products of hot combustion (soot).

Blanket Application: An application of spray or granules over an entire area rather than only in rows, pockets or as band.

Blending: The process of mixing two or more different products together to obtain desired final product.

Blind tillage: Tillage done after seeding or planting either before emergence or early stage of crop. Extra plants and broad-leaved weeds are uprooted.

Blocking or local control: Refers to grouping of experimental units in such a way that units within a group are as homogenous as possible and between groups the experimental units are as heterogeneous as possible. The group of homogeneous experimental units is called block. In agricultural field experiments, the entire experimental units are divided into compact blocks by making use of greater homogeneity of adjacent areas in a field. It helps in reducing the experimental error, thereby increasing the precision of the experiment.

Blue Green Algae (BGA): A heterogenous group of prokaryotic photosynthetic nitrogen fixing organisms which contain chlorophyll 'a'. They include unicellular, colonial and filamentous species. They are obligate phototrophs and store cyanophycean starch.

Blue vitriol: Copper sulphate ($CuSO_4.5H_2$) is also known as blue vitriol. It is used as algaecide as well as potato dehaulming agent at 1.0 ppm.

Bog Soil: An intra-zonal group of silos with a peaty surface underlain by peat; developed under swamp or marsh type of vegetation; mostly found in humid or sub humid regions.

Bolting or Shooting: Refers to significant shoot elongation that precedes flowering in many plants. Example: bolting in onion.

Boltzmann's Constant: It is the universal constant of proportionality between the radiant emittance of a black body and the fourth power of the body's absolute temperature.

Bordeaux mixture: Bordeaux mixture is a mixture of copper (II) sulphate and slaked lime used as a fungicide.

Boro: Rice sown in winter and harvested in summer. This is also called spring rice.

Bowen ratio: The ratio of energy flux upward as sensible heat to latent energy flux in the same direction.

Break-Even Point: Increase in crop yield that is just sufficient to offset fertilizer costs. A no profit – no loss point.

Breeder Seed: Breeder seed is seed or vegetative propagating material directly controlled by the originating or sponsoring plant breeder of the breeding programme or institution and/or seed whose production is personally supervised by a qualified plant breeder and which provides the source for the initial and recurring increase of foundation seed.

Brey's Nutrient Mobility Concept: As the mobility of a nutrient in the soil decreases, the amount of that nutrient needed in the soil to produce a maximum yield increase.

Brix Value: Brix is the sugar content of an aqueous solution. One-degree Brix is 1 gram of sucrose in 100 grams of solution and represents the strength of the solution as percentage by mass.

Broad Bed Furrow (BBF) System: A land management system for successful crop production in vertisols (heavy black soils) by providing better drainage through raised broad beds (above 90-100 cm wide) and furrow (25-30 cm wide).

Broomrape/Orobanche: The name "broomrape" has been derived from the association of certain species of *Orobanche* with "broom" (*Cytisus* sp.) grown in Europe. *Orobanche* sp. Is an obligate, total-/holo-root, phanerogamic plant parasite. It depends entirely on autotrophic plants for its germination and nourishment.

Brownian Movement: Minute particles are in a state of constant motion, as all the particles carry like electric charges, repel one another and recoil on contacting the other particles in the system- called brownian movement.

Buffer Strip: A narrow strip of grass or other close-growing vegetation that, when placed along the contour on a slope, traps sediment that was produced on the hillslope above.

Bulk Density: The mass of dry soil per unit bulk volume. The value is expressed as Mg per cubic meter (Mg m^{-3}).

C:N Ratio: The ratio between organic C and N in organic materials or soils when C:N ratio of crop residue below 20 there is a net gain of mineral nitrogen, when it is above 30 there is a net immobilisation of mineral nitrogen and the crop may even face a temporary nitrogen deficiency. If the C:N ratio is between 20 and 30, there is no mineralisation or immobilisation.

C:P Ratio: Ratio between the organic C and phosphorus in organic materials, soil *etc.* net mineralisation of organic phosphorus results if the C:P ratio is less than 200, and there may be net immobilisation of P above C:P ratio of 300. Application of P biofertilizer improves the Carbon and Phosphorus ratio.

C:S Ratio: The ratio of organic C to sulphur in organic materials, soils *etc.* At C:S ratio of less than 200, net mineralisation of organic sulphur is expected in the soil, while a C:S ratio higher than 400 favours net immobilisation of mineral S.

Cablegation: A method of surface irrigation that uses gated pipe to both transmit and distribute water to furrows or broad strips.

Calcareous Soil: Soil containing sufficient free calcium carbonate or magnesium carbonate to effervesce visibly when treated with cold 0.1 N hydrochloric acid. High content of lime (up to about 5 per cent), particularly in the clay fraction, appreciably increases erodibility by wind.

Calcification: Calcification is the general process by which calcium carbonate ($CaCO_3$) accumulates in soils.

Caliche: A layer near the surface, more or less cemented by secondary carbonates of Ca and Mg precipitated from soil solution.

CAM plants: Crassulacean acid metabolism, also known as CAM photosynthesis, is a carbon fixation pathway that evolved in some plants as an adaptation to arid conditions. In a plant using full CAM, the stomata in the leaves remain shut during the day to reduce evapotranspiration, but open at night to collect carbon dioxide. Example pineapple and cactus plant.

Canopy: The vertical projection downward of the aerial portion of plants, usually expressed as percent of ground so occupied.

Capillary Rise: The rise of water above the hydrostatic surface through the influence of capillarity.

Capillary Water: Water which remains in the soil pore spaces after gravity drainage has occurred. This water resides in the soil pores were capillary forces balance gravity forces such that a negligible amount of further drainage takes place. This water is available for plant use and exist in the soil under suctions in the range of 0.1-31 bars.

Carbon Cycle: The sequence of transformations whereby carbon dioxide is converted to organic forms by photosynthesis or chemosynthesis, recycled through the biosphere (with partial incorporation into sediments), and ultimately returned to its original state through respiration or combustion.

Carbon Dioxide Equivalent (CO_2e): It is the international metric standard for measuring and reporting GHGs, is used to normalize the global warming potentials of different GHGs so that they can be readily compared. *e.g.* CH_4, N_2O and CFC have 25, 298 and 10,900 times GWP than CO_2, respectively, when considered over a 100 years' time horizon.

Carbon Farming or Growing Carbon: The process of soil carbon sequestration is termed carbon farming.

Carbon Footprint: A representation of the effect of human activities on the climate in terms of the total amount of green house gases produced.

Carbon Sequestration: Process through which agricultural and forestry practices removes co_2 from the atmosphere.

Carbon -Nitrogen Ratio (C:N): The ratio of percent C to percent N, or the C:N ratio, defines the relative quantities of these elements in residues and living tissues. Whether nitrogen is mineralized or immobilized depends on the C:N ratio of the organic matter being decomposed by soil microorganisms.

- ✰ Wide C:N ratios of > 30:1-Immobilization of soil nitrogen will be favoured. Residues with wide C:N ratios include hay, cornstalks, dry leaves, and sawdust.
- ✰ C:N ratios of 20:1 to 30:1- Immobilization and mineralization will be nearly equal.
- ✰ Narrow C:N ratios of <20:1-Favour rapid mineralization of N. Residues with narrow C:N ratios include alfalfa, clover, manures, biosolids, and immature grasses.

Cardinal temperature: Below a certain minimum and beyond maximum value of temperatures, plant growth ceases. Between, these limits, there is an optimum temperature at which growth proceeds with greatest rapidity. These three points are known as cardinal temperature.

Carrier: The liquid or solid material added to a chemical compound to facilitate its storage, shipment and application.

Carry Over Effect: The efficient of fertilizer application expressed on a crop that follows the directly fertilized crop. Important for phosphorus and micronutrient management.

Cash Crop Farm: Nutrients come to a modern cash crop farm in fertilizers and other materials applied directly to the fields. Crops harvested from the fields remove a fraction of the applied nutrients, which leave the farm when the crops are sold. On a cash crop farm, there is a direct connection between the flow of nutrients and the agronomic or economic performance of the farm. Traditional economic and agronomic incentives can be effective in guiding nutrient use on cash crop farms to optimize both crop production and environmental protection.

Catchment Area: An area from where entire rainwater is collected and conducted to a reservoir.

Catchment: An independent unit of treated or untreated land area contributing runoff water to a common point.

Catena: A sequence of soils of about the same age, derived from similar parent material, and occurring under similar climatic condition, but having different characteristics due to variation in relief and in drainage.

Cation Exchange: Clays and organic matter typically possess net negative charge. Positively-charged cations are attracted to these negatively-charged particles, just as opposite poles of magnets attract one another. Cation exchange is the ability of soil clays and organic matter to adsorb and exchange cations with those in soil solution (water in soil pore space).

Cation Exchange Capacity: The sum total of exchangeable cations adsorbed by a soil, expressed in milliequivalents per 100 g (c mole) +)/kg) of soil.

Cation: An ion carrying positive charges of electricity. The common soil cations are calcium, magnesium, sodium, potassium and hydrogen.

Certified Seed: Certified seed shall be the progeny of foundation seed and its production shall be so handled as to maintain specific genetic identity and purity according to the standards prescribed for the crop being certified.

Char: A porous solid residue resulting from the incomplete combustion of biomass.

Characteristics Curve or Moisture Release Curve: It is the relationship between soil moisture tension and the moisture content in the range from field capacity to wilting point.

Chelate: An organic compound capable of holding the plant nutrient in a form which prevents it from getting tied with other elements in the soil, thus keeping more or less in available form for the plant. The term refers to the claws of a crab illustrative of the way in which the autumn is held, *e.g.* EDTA.

Chemigation: The process of applying chemicals to crop or soil through an irrigation system with the water.

Chemical Oxygen Demand: A measure of oxygen consuming capacity of inorganic and organic matter present in water or waste water.

Chemotropism: A growth movement in response to a chemical.

Chilling injury: An injury caused by relatively low temperatures above the freezing point.

Chisel: A chisel is a tool with a characteristically shaped cutting edge of blade on its end, for carving or cutting a hard material such as wood, stone, or metal by hand, struck with a mallet, or mechanical power.

Citric acid cycle or Kreb's cycle: A series of cyclic reactions which transforms pyruvic acid into carbon dioxide and hydrogen which then undergoes terminal oxidation to yield available energy.

Clay (soil): Soil material that contains 40 per cent or more clay less than 45 per cent sand and less than 40 per cent silt.

Climate Change: Long term change in the statistical distribution of weather patterns over period of time that range from decades to millions of years.

Climate Resilient Agriculture (CRA): The incorporation of adaptation, mitigation and other practices in agriculture which increases the capacity of the system to respond to various climate related disturbances by resisting damage and recovering quickly.

Climate Smart Agriculture: Promotes production system that sustainably increase productivity, resilience (adoption), reduces/removes GHGs (mitigation), and enhances achievement of national food security and development goals.

Climate: The generalized weather or summation of weather conditions over a given region during comparatively longer period.

Climatic Erosivity: The relative influence of climate on field erodibility by wind in different regions, specifically the effects of average wind speed and effective soil surface moisture.

Climatology: Climatology or climate science is the scientific study of climate, scientifically defined as weather conditions averaged over a period of time.

Clod: A compact, coherent mass of soil greater than 2 millimetres in equivalent diameter, often created by tillage or other mechanical disturbance of the soil.

Cloud Cover: Cloud cover refers to the fraction of the sky obscured by clouds when observed from a particular location. Okta is the usual unit of measurement of the cloud cover.

Coefficient of variation: It is a measure of relative dispersion and is the standard deviation expressed as a percentage of mean.

It is given by: CV (per cent) = (Standard deviation/mean) *100. It is unit free.

Cohort: Seeds in most cases germinate in flushes/batches and seedlings of one flush/lot is called cohort.

Coleoptile: It is a shoot apex encased in sheath.

Colloid: A colloid has gel-like thickened appearance with solute particle sizes ($>1*10^{-4}$ mm) which are much greater than pore diameters of the intermolecular spaces of water. As a result, they form neither solution nor suspension, but a different mixture called colloid. Gel formulation is an example of colloid.

Combo Happy Seeder: A compact, light weight, tractor mounted machine having capability to manage the loose straw and anchored rice residues in strips just in front of each furrow opener.

Command Area: The irrigable land area by the canal or from a dam or an irrigation project.

Companion Crop: It's an intercropping system where the production of both intercrops is equal to than its solid planting.

Competition Index: It is a measure to find out the yield of various crops when grown together as well as separately. In indicates the yield per plant of different crops in mixture and their respective pure stand on a unit area basis.

Complete fertilizer: Having all the three primary nutrients *viz.* Nitrogen, Phosphorus and Potassium.

Completely Randomized Design: A simple form of experimental design which is used when all experimental units are homogeneous and in which the treatments are allocated to the experimental units randomly.

Compound Fertilizer: Such fertilizer has a declarable content of at least two of the major nutrients obtained chemically and generally granular form, *e.g.,* Nitro Phosphate, Ammonium Phosphate and Di ammonium Phosphate.

Concurrent/Direct Allelopathy: It refers to the instantaneous direct effect of released toxins from the living plants to another growing in vicinity. It is differently known as **"live plant effect".**

Conjugation: It refers to aggregation or binding-up of herbicides with several parent compounds or their metabolites in plants such as amino acids, peptides, lignins, lipids, carbohydrates, sugars/glucose, proteins and formation of esters, ethers, amides or anilides as conjugated products. It is a pathway of herbicide transformation in many plants.

Conjunctive Use of Water: It refers to the use of saline water of wells in combination with canal water.

Conservation Agriculture: The concept of resource saving agricultural crop production that strives to achieve acceptable profit together with high and sustained production levels while concurrently conserving the environment.

Conservation Tillage: Any tillage or planting system in which at least 30 per cent of the soil surface is covered by plant residue after planting to reduce soil erosion.

Consumptive Irrigation Requirement: It is the depth of irrigation water exclusive of precipitation, stored soil moisture or groundwater that is required to meet evapo-transpiration demand of the crop.

Consumptive Use: Quality of surface and groundwater absorbed by the crop and transpired or used directly in the building of plant tissues, together with that evaporated from the cropped area, expressed in units of volume per unit area. It is the sum of total water transpired and used for carrying out metabolic activities and building-up the plant body by vegetation and water evaporated from unit area of vegetation. It is expressed in cm depth of water per unit area.

Contact Herbicide: Contact herbicides kill only the parts of the plant they touch.

Contingency Planning: Contingency crop planning refers to implementing a plan for making alternate crop or cultivar choices in tune with the actual rainfall situation and soils in a given location.

Contingent Drought: Abnormal failure of rainfall. It occurs almost anywhere especially in most parts of humid and sub humid climates.

Contour Bunding: The construction of small bund across slope of the land on a contour, so that the long slope is cut into a series of small ones and each contours bund act as a barrier to the flow of water.

Contour Farming: The practice of using ridges and furrows left by tillage to redirect runoff from a path directly downslope to a path around the hillslope.

Control Plot: A standard check plot used for comparing the relative performance of treatments under study.

Convection: Processes of transmission of heat through actual movement of molecules of the medium.

Conveyance Efficiency (Ec): The percentage of water introduced to the farm at the inlet structure to that actually reached the field to be irrigated. Losses may occur due to seepage, evaporation, leaks, *etc.*

$$= \frac{\text{Water delivered to field}}{\text{Water diverted from source}} \times 100$$

Cooking coefficient (CC): This is a ratio of difference between length of cooked and raw grain to that of difference between breadth of cooked and raw grain.

Correlation Coefficient (r): Refers to the measure of the strength of a linear association between two quantitative variables. It is unit free and lies between -1 to +1. A correlation coefficient of 0 means that the two variables are not linearly related. An 'r' value of -1 or +1 indicates complete linear association. A positive correlation means that high values of one variable are associated with high values of the other. A negative correlation means that as one variable increases the other tends to decrease.

Corrugation Irrigation: A method of surface irrigation similar to furrow irrigation, in which small channels, called corrugations, are used to guide water across a field.

Corrugation: A method of surface *irrigation* similar to furrow *irrigation*, in which small channels, called *corrugations*, are used to guide water across a field.

Covariance: It is a measure of how much two variables change together. If two variables tend to vary together (that is, when one of them is above its expected value too), then the covariance between the two variables will be positive. On the other hand, when one of them is above its expected value the other variable tends to be below its expected value, then the covariance between the two variables will be negative. It is computed as the mean f the product of corresponding deviation o two variables from their individual means. From a sample data, it is estimated as the ratio of the sum of cross products to the corresponding number of degrees of freedom.

Cover Crop: Close-growing crop that provides soil protection, seeding protection and soil improvement between periods of normal crop production, or between trees in orchards and vines in vineyards. When incorporated into the soil, cover crops may be referred to as green manure crops.

Critical Difference: The standard error of difference of means when multiplied by 2 and t-value (at required level of significance) at the error degrees of freedom is known as critical difference. It is also known as least significant difference (LSD) or Fisher's Protected LSD. Also refer to standard error of difference of means and Least Significant Difference test.

Critical Growth Period: Plants must have ample moisture throughout the growing season for optimum production and the most efficient use of water. This is most important during critical periods of growth and development. Most crops are sensitive to water stress during one or more critical growth periods in their growing season. Moisture stress during a critical period can cause an irreversible loss of yield or product quality.

Critical Period (water): It is the period during which a crop is affected severely due to moisture stress and the loss cannot be compensated by adequate moisture supply in other periods or stages.

Critical Value: A value of a diagnostic factor which forms the boundary between suitability rating classes.

Critical period or critical stages (crop): The period or the stage of development in the fife-cycle of a crop or a plant when it is most sensitive to the deficiency of a production factor and is most responsive to the correction of the deficiency.

Crop and Livestock Farm: On farms with livestock (*e.g.*, dairy), a large proportion of the plant nutrients from crops produced as feed for the animals are traditionally returned to the farm fields in manure.

Crop coefficient: Crop coefficients are properties of plants used in predicting evapotranspiration (ET). The most basic crop coefficient, K_c, is simply the ratio of ET observed for the crop studied over that observed for the well calibrated reference crop under the same conditions.

Crop Ecology: Crop ecology may be defined as the study of crop plants in relation to their physiological environment.

Crop Evapotranspiration (ETc): Crop evapotranspiration sometimes called crop consumptive use, is the amount of water that plants use in transpiration and building cell tissue plus water evaporated from an adjacent soil surface. Seasonal local crop water use requirements are essential for planning crop production systems.

Crop Growth Rate (CGR): It represents dry weight gained by a unit area of crop in a time expressed as g day^{-1} m^{-2}. It is given as

$$= \frac{W_2 - W_1}{t_2 - t_1}$$

Where W_1 and W_2 are plant dry weight at time t_1 and t_2, respectively.

Crop Logging: Foliar diagnosis comparing the nutrient status of comparable leaves of high and low yielding crop plants (generally used in sugarcane in Hawaii). This was proposed by Clements.

Crop Models: The only automated source of dynamic knowledge within computer technology or any other known technology that can respond to crop status, current meteorology, and management practices.

Crop Root Zone: The soil depth from which a mature crop extracts most of the water needed for evapotranspiration. The crop root zone is equal to effective rooting depth and is expressed as depth in inches or feet.

Crop rotation: A crop rotation is a sequence of different crops grown in a recurrent sequence over a given number of years. In some rotations a crop may occupy the land two years in succession. To realize the greatest benefits, a crop rotation should not have the same annual crop grown 2 years in succession and should alternate plant families. This minimizes the potential for build-up and carryover of insect and disease populations, and maintains some degree of.

Crop Water Productivity: It is the amount of water required per unit of yield, and is a vital parameter to access the performance of irrigated and rainfed agriculture. It is crop yield per unit water consumptively used in evapotranspiration and can also be expressed in terms of wet or dry yield, nutritional value or economic return.

Crop Water Use Efficiency: The ratio of crop yield to the amount of water depleted by the crop plants through evapotranspiration.

$$= \frac{\text{Yield (kg/ha)}}{\text{Evapotranspiration (mm)}}$$

Crop Weather Calendar: The crop weather calendar consists of three parts *viz.*, crop husbandry, climate requirements and weather warnings. These crop weather calendars help the weather forecaster to see at a glance the type of

weather warnings are to be issued for a particular district in a given weather situation during a particular phase of crop.

Crop Yield Index: A measure of comparison of the yields of all crops on a given farm with the average yield of these crops in the locality. The relationship is expressed in percentage.

Crop-Associated Weeds: They are weeds which are crop-specific not by virtue of parasitism but by certain other reasons, such as specific micro-climate requirement, mimicry with crops *etc., e.g. Chicorium intybus* and *Coronopus didymus* in berseem.

Crop-Bound Weeds: They are weeds which usually parasitize a crop for their food and survival partially or wholly, *e.g. Cuscuta* sp., *Orobanche* sp., *Striga* sp., *Loranthus longiflorus* and *Cassytha filiformis.*

Cropping index: The number of crops grown per annum on a given area and actual net cultivated area expressed in percentage.

$$= \frac{\text{Total cropped area in a year}}{\text{Net cultivated area}} \times 100$$

It does not consider the length of growing period in crops.

Cropping Intensity: Cropping intensity refers to raising of a number of crops from the same field during one agricultural year; it can be expressed through a formula. Cropping Intensity = (Gross Cropped Area/Net Sown Area) x 100.

Cropping Pattern: Yearly sequential and spatial arrangement of crops, and crops and fallow on a given area is referred as cropping pattern (Usually representing larger landscape like district, delta region and state).

Cropping Scheme: It is a plan according to which crops are grown on individual plots of a farm during a given period of time with the object of obtaining maximum return from each crop without impairing the soil fertility.

Cropping System: The cropping patterns used on a farm and their interaction with farm resources, other farm enterprises and available technology, which determine their make up.

Crown: The point were stem and root join in a seed plant.

Crust: A thin surface layer, where aggregates are bound together and the surface is sealed. It is more compact and mechanically stable than the soil material immediately beneath it. Crust is characterized by its dense, platy structure that becomes less distinct with depth until it merges with the soil below. Crust is a transitory condition. Deposition The accumulation of eroded soil material on the land surface when the velocity of the transporting agent (wind or water) is reduced.

Culm: Joined stem made up of nodes and internodes in alternate order.

Cultivated Area: Cultivated area is the land area that is cropped every season for seasonal crops (rice, maize, wheat, soybeans, *etc.*) or every year for longer season crops like sugarcane or perennial crops like fruit trees.

Cultivated land-utilization index: It is calculated by summing the products of land area planted to each crop, multiplied by the actual duration of that crop and divided by the total cultivated land area times 365days.

Cultivation index: The ratio of organic carbon in cultivated soil to that in virgin soil.

Curing: Tobacco curing is also known as colour curing, because tobacco leaves are cured with the intention of changing their colour and reducing their chlorophyll content.

Dapog Rice Nursery: Rice seedlings are raised on well prepared soil either in field or in nursery boxes, covered under banana leaves of plastic with bamboo pegs. Dapog seedlings are ready for transplanting in 9 to 14 days after the seeds are sown. It is labour and time saving method.

Deep tillage or sub-soiling: Refers to tillage that loosens the soil with a purpose of conserving in the monsoon season or for breaking hard pans. (40 cm deep or more)

Darcy's Law: The law stating that the velocity of a fluid in permeable media is directly proportional to the hydraulic gradient. In this law, hydraulic conductivity (K) is taken as proportionality constants. Darcy stated that the rate of flow increases with an increased depth of water above the bottom of the soil, and decrease with an increased depth of soil, through which water flows.

Data: Data refers to a collection of organized information, usually the result of experience, observation or experiment.

Dead Furrow: At the time of sowing or immediately after sowing, deep furrows of 20 cm depth are formed at intervals of 6-8 rows of crops. No crops are raised in this furrow.

Decomposition: The breakdown of mineral and organic materials.

Deep tillage or sub-soiling: The subsoiler is a tillage tool which will improve growth in all crops where soil compaction is a problem. In agriculture angled wings are used to lift and shatter the hardpan that builds up due to compaction. The design provides deep tillage, loosening soil deeper than a tiller or plough is capable of reaching.

Deficit Irrigation: Irrigation water management alternative where the soil in the plant root zone is not refilled to field capacity in all or part of the field.

Deflocculated: To separate or break down soil aggregates of clay into their individual particles.

Defoliant: A defoliant is any herbicidal chemical sprayed or dusted on plants to cause their leaves to fall off.

Delinting: Delinting is the process of removal of fuzz from the seed coat in cotton *i.e.* it is a crop specific seed management technique.

Delta (Water): The total depth of water required by a crop.

Design of Experiment: Allocation of treatments to different experimental units to generate meaningful information provided all other conditions except treatments are same for all experimental units. The effects of uncontrolled variations in fertility and other natural factors are minimized.

Desilting: The removal of excess silt and mud from farm ponds and reservoirs and usually carried out when water levels are low.

Dessicant: A desiccant is a hygroscopic substance that is used to induce or sustain a state of dryness in its vicinity.

Desuckering: Removal of surplus and unwanted suckers from banana plant is known as desuckering. Suckers are removed from the mother either by cutting the sucker at ground level or by destroying the heart of the suckers without detaching the sucker from the plant.

Detachment: The removal of transportable fragments of soil material from the soil mass by an eroding agent, usually falling raindrops, running water, wind, or windblown soil particles. Detachment is the process that makes soil particles or aggregates available for transport.

Detasseling: Removing the immature pollen-producing bodies, the tassel, from the tops of corn plants and placing them on the ground. It is a form of pollination control, employed to cross-breed, or hybridize, two varieties of corn.

Detergent: A cleaning substance such as soap. Detergents are surface active agents (surfactants), many having applications in pesticide formulation as emulsifiers and wetting agents.

Diffusion: Diffusion is the movement of a substance from an area of high concentration to an area of low concentration.

Dispersal: Movement of seeds and other plant propagules to short or long distance, away from the mother plant.

Diversity Index: It measures the multiplicity of crops or farm products which are planted in a year by computing the reciprocal of sum of squares of the share gross revenue received from each individual farm enterprise in a single year.

Dockage: Percent impurity in a seed sample.

Dolomite: A natural mineral composed of calcium and magnesium carbonate, widely used as a liming material and as an ingredient in fertilizer mixtures. Dolomite is used in fertilizer to render them non-acid forming. It also supplies available magnesium.

Dormancy: Dormancy is defined as the temporary inability of a viable seed to germinate under favourable environmental conditions.

Drainage: Drainage is the natural or artificial removal of a surface's water and sub-surface water from an area with excess of water.

Drip Irrigation: This is an irrigation method which minimize the use of water and fertilizer by allowing water to drip slowly to the roots of the plants, either onto the soil surface or directly onto the root zone, through a network of valves, pipes, tubing, and emitters. This is a trickle irrigation or micro irrigation.

Drizzle: Drizzle is a light liquid precipitation consisting of liquid water drops smaller than those of rain, generally smaller than 0.5 mm (0.02 inch) in diameter. Drizzle is normally produced by low stratiform clouds and stratocumulus clouds.

Drosometer: An instrument for measuring the amount of dew formed on a given surface.

Drought Year: Any year when precipitation is less than 80 per cent of the long-term normal.

Drought: A drought is an event of prolonged shortages in the water supply, whether atmospheric, surface water or ground water. A drought can last for months or years or may be declared after as few as 15 days.

Dry Aggregate: A compound or secondary air-dry soil particle that is not destroyed by dry sieving.

Dry Farming: Cultivation of crops in areas where rainfall is less than 750 mm per annum

Dryland Farming: Crop production without irrigation (rainfed agriculture).

Duckage: Dockage is defined as material intermixed with a parcel of grain that can be separated by normal cleaning procedures.

Dust Storm: A strong turbulent wind carrying large amounts of soil particles in suspension.

Duty of Water: The total volume of irrigation water required for a particular type of crop to mature. It includes consumptive use, evaporation and seepage from ditches and canals, and the water eventually returned to streams by percolation and surface runoff.

$$= \frac{8.64 \times \text{Base}}{\text{Delta}}$$

Duty: It is the area irrigated by one cusec discharge of water during the crop period. It is equal to twice the base divided by delta.

Dwarfing gene: The dwarfing genes reduce plant height by altering the sensitivity to growth-related hormones or the molecular pathway involved.

Ear head Malformation: It is a formative effect or physiological deformity in the ear head of some varieties of wheat like 'Kundan' and 'Arjun' ('HD 2009') induced by 2,4-D application.

Earthing up: Earthing up or ridging is the technique in agriculture and horticulture of piling soil up around the base of a plant. It can be done by hand (usually using a hoe) or with powered machinery, typically a tractor attachment.

Ecesis: The process by which a plant or animal becomes established in a new habitat.

Ecology: Study of mutual relationship between organisms and their environment.

Ecophene: It is a non-heritable morphological adaptation in plants to prevailing environmental stresses.

Ecosystem: An ecosystem has been defined as an open system comprising plants, animals, organic residues, atmospheric gases, watered minerals that are involved together in the flow of energy and circulation of matter.

Ecotypes: Ecological races in which the variations associated with certain factors of environment are genetically fixed.

Edaphic: A term pertaining to the influence of relationship of soil or other similar media to plant growth in contrast to atmosphere influences.

Edaphology: Edaphology is concerned with the influence of soils on living things, particularly plants. Edaphology includes the study of how soil influences humankind's use of land for plant growth as well as people's overall use of the land.

Effective Irrigation: A controlled and uniform application of water to crop plant in required amount at required time, with minimum cost to produce optimum yields without waste of water and adverse effect on soil in the form of soil salinity and water-logging problems.

Effective Precipitation: That portion of the total rainfall precipitation which becomes available for plant growth.

Effective Rainfall: The fraction of total precipitation that forms the soil water reserve for consumptive use of crop or that becomes available for crop production.

***El Nino*/Southern Oscillation (ENSO):** It is a band of warm ocean water temperature that periodically develops off the pacific coast of south America. Typically, this anomaly happens at irregular intervals of two to seven years, and last nine months to two years. The average period length is five years. When this warming occurs for only seven to nine months, it is classified as *El Nino* "conditions"; when it occurs for more than the period it is classified as *El Nino* "episodes". *El Nino* is Spanish for "The Boy", and the capitalized term *El Nino* refers to the Christ child, Jesus, because periodic warming in pacific near south America is noticed around Christmas.

Elasticity of Production: The percentage increase in output as compared to the percentage increase in input.

Electricity conductivity (EC): The electrical conductance of an extract from a soil saturated with distilled water, normally expressed in units of decisiemens per meter at 25°C.

Electro-magnetic spectrum of light: The electromagnetic spectrum is the range of frequencies of electromagnetic radiation and their respective wavelengths and photon energies.

Element: Substances which consists all matter at and above atomic level. An element consists of atoms of same atomic number.

Emulsifiable Concentrates (EC): An emulsified concentrate is an oil-based liquid compound produced by dissolving the toxicant and an emulsifying agent in an organic solvent. They are usually clear solution and form emulsion when added to water before spraying and the resultant turns milky white or other colours (yellow in case of dinitroanilines). Pendimethalin, fluchloralin, paraquat, 2,4-DEE, chloroacetamides, carbamates and thiocarbamates have EC formulation.

Emulsion: A dispersion of fine particles of oily material in water (OW type) or less commonly in pesticide usage a dispersion of water droplets in oil (WO type). It is a heterogeneous mixture/medium of two liquids (stable or unstable) held together by a third chemical called emulsifier or emulsifying agent, which helps to mix active ingredient with water. In emulsion, the component compounds ingredient with water. In emulsion, the component compounds maintain their original identity all through. Milk is a good emulsion.

Endotrophic (mycorrhiza): An association between fungi and plants were in fungi live intracellularly for most part of their life.

Energy Canes: Energy cane is sugar cane that has been genetically modified to become more productive in the manufacture of biofuel, biochemicals and renewable energy generation.

Enforced Dormancy: It is the dormancy in weed seeds caused by the absence of germination factors, *e.g.* temperature, moisture, air (CO_2 and O_2) and light (particularly red light).

Enriched Compost: Compost fortified with the addition of fertilizers during its production to raise its nutrient content and narrow down the C:N ratio.

Enrichment (soil): General term for addition of nutrient material to a soil body.

Entisols: Soils that have no diagnostic pedogenic horizons.

Eolian: Soil transported by wind.

Ephemeral Gully Erosion: Erosion that occurs from the action of runoff water which concentrates in shallow flow channels when rills converge. These flow channels are alternately filled with soil by tillage operations and re-formed in the same general location by subsequent runoff events.

Ephemeral Weeds: Weeds which just grow and complete their life cycle within a very short period are ephemeral weeds, *e.g. Ageratum conyzoides* and *Phyllanthus* sp. It is easy to control them but they may be as harmful as other weeds to crops.

Ephiphytes: An epiphyte is an organism that grows on the surface of a plant and derives its moisture and nutrients from the air, rain, water (in marine environments) or from debris accumulating around it.

Epigeal: In dicotyledones, both cotyledons come out entirely or partially at the time of germination, called epigeal germination.

Epinasty/Epinastic symptoms: Epinasty means bending, twisting or curling of leaves, stems or even ear head due to herbicide, which later become brittle and break up by light pressure. Epinastic effects are usually associated with altered cell division, cell elongation and tissue differentiation and induced mainly by hormone herbicides *e.g.* phenoxy-alkanoic acids. The increased growth on the upper surface of a plant organ or part (especially leaves) causing it to bend downward.

Equivalent Acidity: The number of parts by weight of $CaCO_3$ required to neutralise the acidity results from the use of 100 parts by weight of fertilizer.

Equivalent Basicity: The number of parts by weight of $CaCO_3$ corresponds in acid neutralising capacity to 100 parts by weight of the fertilizer.

Erodibility: The susceptibility of soil to erode. Soils with low erodibility include fine textured soils high in clay that are resistant to detachment, and coarse textured soils high in sand that have low runoff. Soils having a high silt content are highly susceptible to erosion.

Erosivity: The energy (amount) and intensity of rainstorms that cause soil to erode. An erosivity includes the effects of raindrop impact on the soil and the amount and rate of runoff likely to be associated with the rain.

Error Mean Square: The residual or error sum of squares divided by the number for degrees of freedom on which the sum is based. It provides an unbiased estimate of error variance.

Erucic Acid: The C_{22} fatty acids with a double bond; a homologue of oleic acid with four more carbons. Oils with a high erucic acid are considered as unsuitable for human consumption.

Essential Element or Nutrients: An essential mineral element is one that is required for normal plant growth and reproduction. With the exception of carbon (C) and oxygen (O), which are supplied from the atmosphere, the essential elements are obtained from the soil. The amount of each element required by the plant varies; however, all essential elements are equally important in terms of plant physiological processes and plant growth.

Eutrophication: A process that increases the amount of nutrients, especially nitrogen and phosphorus, in a marine or aquatic ecosystem. Eutrophication occurs naturally over geological time but may be accelerated by human activities, such as waste disposal or land drainage, leading to an increase in algae and a decrease in diversity.

E-value or Et value: It is a measure of the amount of the isotopically exchangeable phosphate of the soil and like L-value, the E-value measures labile P, made up of 'surface-p' and 'solution-p'. Ti is the laboratory 'L' value.

Evaporation: Evaporation is the process whereby liquid water is converted to water vapor (vaporization) and removed from the evaporating surface. Energy is required to change the state of the molecules of water from liquid to vapor. Direct solar radiation and, to a lesser extent, the ambient temperature of the air provide this energy. As evaporation proceeds, the surrounding air becomes gradually saturated and the process will slow down and might stop if the wet air is not transferred to the atmosphere. The replacement of the saturated air with drier air depends greatly on wind speed. Hence, solar radiation, air temperature, air humidity, and wind speed are climatological parameters to consider when assessing the evaporation process.

Evapo-Transpiration: Evapo-transpiration or consumptive use denotes the quantity of water transpired by plants during their growth or retained in the plant tissue, plus the moisture evaporated from the surface of the soil and the vegetation. It is the total amount of water lost due to the transpiration by a crop and evaporation from the soil surface during a specified time from a particular area.

Exchangeable Cation Percentage (ECP): This term indicates the degree of saturation of soil or exchange capacity with the cation and is expressed as follows

$$= \frac{\text{Exchangable cation (meg/100 g soil)}}{\text{Cation exchange capacity (meg/100 g soil)}} \times 100$$

Exchangeable Phosphate: The phosphate ion reversibly attached on the surface of the solid phase of soil, in such form that it may go into molecular solution by anionic equilibrium reaction with other anion of the solution phase.

Exchangeable Potassium: The potassium which is held mainly by the colloidal portion of the soil and is easily exchanged with the anion of the natural salt solution less the water-soluble potassium. It is readily available to growing plant roots.

Exchangeable Sodium Percentage: The degree of saturation of the soil exchange complex with sodium expressed in percentage.

$$= \frac{\text{Exchangable Na (meg/100 g soil)}}{\text{Cation exchange capacity (meg/100 g soil)}} \times 100$$

Exosphere: The outer most layer of the earth's atmosphere is named as the exosphere and this layer lies between 400 and 1000 km.

Exotic/Introduced Weeds: Weeds originating in a country, but got introduced to another country through process of dispersal in course of time are exotic/introduced weeds to that country, *e.g. Sorghum halepense* was introduced intentionally to India from USA as forage crop, *Opuntia stricta* from Australia as hedge crop, and *Mikania micrantha* from Malaysia.

Experiment: A systematic procedure for making observations under controlled conditions in such a way that they can be used for arriving at general conclusions regarding the population under study.

Experimental Design: A logical structure of an experiment, that helps in obtaining results with some precision.

Factor: A factor is an input or process or controllable variable which is varied in the experiment at the will and under the control of the experimenter. (Also refer to level of a factor).

Facultative Weeds/Apophytes: Weeds growing primarily in the wild communities, but often encroach cultivated fields and associate them closely with human's activities, mainly crop cultivation are "facultative weeds". They grow wild in primary habitat as well as in association with crops in the cultivated fields (secondary habitat). They are also called apophytes, *e.g. Anagallis arvensis, Galinsoga parviflora, Bidens Pilosa* and *Amaranthus* sp.

Fallow: The practice of leaving land uncropped, either weed-free or with volunteer vegetation, during at least one period when a crop would normally be grown; done to control weeds, or accumulate water or available plant nutrients.

Family Farming: It is a means of organising agricultural, forestry, fisheries, pastoral and agricultural production which is managed and operated by a family and predominantly reliant on family labour including both men and women.

Farm Budgeting: A process of estimating cost, returns and net profit on a farm

Farm Management: The branch of agricultural economics that deals with the business principles and practice of farming with an object of obtaining the maximum possible return from the farm as a unit under a sound farming programme. It is an example of micro-economics.

Fertile Soil: A soil with adequate amount of essential plant nutrients in available form to support optimum crop production.

Fertility: The quality of a soil that enables it to provide nutrients in adequate amounts and in proper balance for the growth of specified plants or crops.

Fertilizer: Any organic or inorganic material of natural or synthetic origin (other than liming materials) that is added to a soil to supply one or more plant nutrients essential to the growth of plants.

Fertilizer Grade: An expression showing the legal guarantee of its available plant nutrients expressed as percentage of plant nutrients in the fertilizer, *e.g.* a 10 – 5 – 5 grade of fertilizer indicates 10 per cent N, 5 per cent P_2O_5 and 5 per cent K_2O.

Fertilizer Use Efficiency (FUE): A measure of the increase in crop yield obtained per unit fertilizer nutrient applied. Agronomically expressed as unit of yield increased per unit nutrient applied.

Fertilizer Use Intensity: The rate fertilizer used per unit cropped area *i.e.* 100 kg fertilizer added per hectare. Can be expressed on net or gross cropped area basis.

Field Application Efficiency: Ratio of water made directly available to the crop and that received at the field inlet.

Field Capacity: The amount of water held against the force of gravity, typically ranges from 1 to 2 inches in each foot of soil. The finer the particles (silts and clays) the more water the soil holds. Extremely coarse sandy soils are typically unable to store moisture in sufficient quantities for crop growth under dry land systems.

Field Channel Efficiency: Ratio between the water received at the field inlet and that at the inlet of a block of fields.

Field Efficiency: It is the ratio of effective field capacity and theoretical field capacity expressed in percentage.

Field Irrigation Efficiency: It is the ratio expressed in the percentage of irrigation water available for use of crops to tat delivered to the field.

Field Water Use Efficiency: It is the ratio of crop yield to the total amount of water used in the field.

$$= \frac{\text{Yield}}{\text{Water used in the field}}$$

Flaccid: A flaccid plant cell is one in which the plasma membrane is not pressed tightly against the cell wall.

Flag Leaf: The last leaf to emerge is called the Flag Leaf, and it is said to constitute about 75 per cent of the effective leaf area that will feed the head and fill the grain.

Flexible Cropping: Flexible cropping is defined as seeding a crop when stored soil water and rainfall probabilities are favourable for satisfactory yield.

Flocculate: To aggregate individual particles into small groups or granules.

Flooping: It refers to wilting of tobacco due to water logging.

Flux: The amount of an entity received by or transferred across a particular plan per unit time.

Foliar Diagnosis: An estimation of mineral nutrient deficiencies or excess in plants based on examination of the chemical composition of the plant parts, and the colour and growth characteristics of the foliage of the plants.

Food Processing Wastes: Food processing wastes can include peels, tops, trimmings, cull or damaged fruit and vegetables from packing houses and canneries, as well as filter press cakes from facilities that produce fruit or vegetable juice. These materials have a low nutrient content and a low C:N ratio. The primary value of these materials, as with the municipal solid waste materials, is in the organic matter they can add to the soil rather than their nutrient content.

Foot-candle: A unit of illumination (now little used) equal to that given by a source of one candela at a distance of one foot (equivalent to one lumen per square foot or 10.764 lux).

Foundation Seed: Foundation seed shall be the progeny of Breeder seed or be produced from foundation seed, which can be clearly traced to Breeder seed.

Free Air CO_2 Enrichment Technology (FACE): This system provides a realistic, cost-effective method for evaluating the effect of supra-ambient CO_2 concentrations on growth, development, yield and water use of agriculture crops and natural ecosystems without any direct perturbation of microclimate.

Free Air Temperature Enrichment Technology (FATE): FATE components include two 1500-watt IR lamps, which can homogeneously irradiate 40-50 cm path of vegetation at an angle of 40^0 above ground. These lamps are regulated by proportional action controller, which modulates IR flux density to obtain a preset target differential (T) between heated and a corresponding unheated plot. IR lamp, is used in simulating temperature, are made of tungsten filament and can irradiate at 2000^0C. filter is used to selectively cut off visible light so that light can be continued in night. It also removes wavelength to which phytochrome is sensitive so that photomorphogenesis can be avoided. FATE irradiation is uniform in space and time.

Friable Index: The difference between shrinkage and plastic limits, in medium and fine textured soils is the friable index.

F-Test: An F-test is any statistical test in which the test statistic follows F-distribution if the null hypothesis is true. The name was coined by George W. Snedecor, in honour of Sir Ronald A. Fisher. It is an alternative name of the variance ratio test developed by Fisher in 1920. It is used to test the equality of the means of multiple normally distributed populations, all having the same variance, the hypothesis that a proposed regression model fits well, to test the equality of variances of two normally distributed populations.

Fulvic Acid: A term of were in usage but usually referring to a mixture of organic substances remaining in solution upon acidification of a dilute alkali extract of soil.

Functional Allelopathy: A substance may be released as a precursor and may be changed into an active substance to inhibit growth. It is the release into the environment of compounds that are toxic after chemical modification by microorganisms.

Furrow: A trench formed a tool in the soil during operation.

Gas Constant: R = 8.3145 x 10^7 erg mol.k^{-1}.

Geographical Area: A demarcated area of the Earth.

Geological Erosion: It is said to be in equilibrium with the soil forming process. It takes place under natural vegetation cover completely undisturbed by biotic factors.

Geometric mean: It is the nth root of the product of 'n' items or values.

Germplasm: Germplasm is living tissue from which new plants can be grown. It can be a seed or another plant part – a leaf, a piece of stem, pollen or even just a few cells that can be turned into a whole plant.

Ginning percentage: The weight of cotton lint obtained from seed cotton expressed in terms of percentage of seed cotton or percentage of lint obtained from 100 units of seed cotton by weight.

Ginning: Ginning is the process of separating the cotton fibers from the *cotton* seeds.

Gleization: Development of a glei (a Russian word meaning blue, grey or green clay) in the lower part of the soil profile above the parent material due to poor drainage and water logging.

Global Dimming/Global Cooling: Decrease in the amount of solar radiation reaching the surface of the earth.

Global Warming Potential (GWP): Measurement of atmospheric heat-trapping ability of a given GHG expressed in terms of an equivalent amount of CO_2. This means methane is approximately more heat-absorptive than carbon dioxide per unit of weight. Similarly, the GWP of N_2O is 310.

Global Warming: Increase of earth's average surface temperature due to effect of green house gasses.

Golden rice: It is variety of rice *Oryza sativa,* produced through genetic engineering to biosynthesize beta-carotene, a precursor of vitamin A, in the edible part of rice. Golden rice was created by transforming rice with only two beta-carotene biosynthesis genes: phy (phytoene synthesis) from daffodil (*Narcissus pseudonarcissus*) *crtI* (carotene disaster) from the soil bacterium *Erwinia uredovora.*

Goodness of Fit: In general, the goodness of agreement between an observed set of values and a second set which are derived wholly or partly on a hypothetical basis, that is to say, derive from the 'fitting' of a model to the data. The term is used especially in relation to the fitting of theoretical distributions to observations and the fitting of regression lines. The excellence of the fit is often measured by some criterion depending on the squares of differences between observed and theoretical values, and if the criterion has a minimum value the corresponding fit is said to be 'best'.

***Gossypium arboretum*:** A diploid old-world cotton species with n=13 chromosomes, which is widely grown in rainfed areas.

***Gossypium barbadense* (Island cotton or Egyptian cotton):** A tetraploid new world cotton species with n=26 chromosomes.

***Gossypium herbacium*:** A diploid old-world cotton species with n=13 chromosomes, constitutes a fairly large percentage of medium staple cotton grown in India.

***Gossypium hirsutum* (American or Upland cotton):** A tetraploid new world species with n=26 chromosomes.

Gossypol: Gossypol is a natural phenol derived from the cotton plant (genus Gossypium). Gossypol is a phenolic aldehyde that permeates cells and acts as an inhibitor for several dehydrogenase enzymes. It is a yellow pigment.

Grass tetany (hypomagnesemia): Condition of cattle and sheep marked by titanic stagers, convulsions, coma and frequent dead; characterised by a low level of blood magnesium.

Gravitation Water: It is the excess water above field capacity in a soil moving down under the influence of gravity.

Green Leaf Manure: This refers to turning under of green leaf and tender green twigs collected from shrubs and trees grown on bunds, wastelands an nearby forest areas. *e.g.* glyceridia, karanj and subabul.

Green Manure Crop: Any crop grown for soil improvement by being incorporated into the soil while green or soon after maturity.

Green manure *in situ*: A practice of ploughing or turning into the soil undecomposed green manure crops in the same field were the crop is grown.

Green Manuring: A practice of ploughing or turning into the soil undecomposed green plant material for improving the physical condition of the soil or for adding nitrogen when the green manure crop is a legume.

Green Seeker: It is a variable rate application and mapping system designed for on-farm use. Green seeker's ground-based sensors provide real time data, day or night, regardless of weather conditions. The data can be used to make variable rate application, mapping of crop biomass and vigour, creating management zones, identify pest and disease problems, evaluate efficiency of drainage systems, modify soil sampling strategies, monitor and modify irrigation schedule, determine optimum harvesting dates, *etc.*

Greenhouse Effect: The absorption of solar radiant energy by the Earth's surface and its release as heat into the atmosphere; longer infrared heat waves are absorbed by the air, principally by carbon dioxide and water vapor, thus, the atmosphere traps heat much as does the glass in a greenhouse.

Greenhouse Gases: Major gases are CO_2 (55 per cent), chloro-fluoro-carbons (25 per cent), methane (11 per cent), NO_2 and others (4 per cent). Carbon dioxide arises primarily from fossil fuel use, CFCs from industrial applications, and others from several sources. Paddy soils are believed to contribute about: >30 per cent of the methane evolved.

Grid Sampling: Taking soil samples from the field in a fixed pattern.

Gross Cropped Area: Total area sown once as well as more than once in a particular year. When the crop is sown on a piece of land for twice, the area is counted twice in GCA.

Gross Domestic Product: The total value of goods produced and services provided in a country during one year.

Gross Irrigated Area: Gross irrigated area is the total irrigated area under various crops during the year counting the area irrigated under more than one crop during the same year as many times as the number of crops grown, crops sown mixed being taken as one crop.

Gross Return: Total monetary returns of the economic produce. The total return expressed in terms of unit area, usually one hectare.

Groundwater: That portion of the water below the surface of the ground at a pressure equal to or greater than atmospheric.

Growing Degree Day: Sum over the growing season of the crop of the difference between the daily temperatures and a reference temperature

$$= \frac{\sum T_{\max} + T_{\min}}{2} - T_{base}$$

Growth Curve: A graph in which the size of some plant characteristics is plotted against time or age. The rate of growth of any single cell or mass of cells are sigmoid (S shaped), slow at first, increases with time to a maximum and then fall off at about the same rate to zero. The period of maximum growth is called grand period of growth.

Gully Erosion: Removal of soil by running water, with formation of channels, that cannot be smoothed out completely by normal condition.

Gypsum: A hydrated calcium sulphate ($CaSO_4.2H_2O$). the commercial material contains varying amount of impurities and is used as a soil amendment.

Half-life ($T_{½)}$: It is the time by which half or 50 per cent of the pesticide applied will disappear from the scene usually soil. In other words, it is the time by which the activity of a pesticide is reduced by 50 per cent. It is an inherent characteristic of each pesticide/herbicide and varies across them. Higher the half-life of a herbicide, longer is its persistence in soil and *vice-versa*.

Hard Seed: Seed that is dormant due to a seed coat impervious to either water or oxygen.

Harmonic mean: It is the inverse of the arithmetic mean of the inverse of given numbers.

Hay: Hay is grass, legumes, or other herbaceous plants that have been cut and dried to be stored for use as animal fodder.

HCN poisoning: Forages such as sorghum, Sudan grass, lotus sp and white clover contain naturally occurring cyanogenic glucosides. Large amount of HCN maybe released through cyanogenesis processed in a relatively short period of time when plant tissue is injured by cutting or grassing. Ruminants are considered susceptible to HCN poisoning because rumen microflora are also capable of enzymatic breakdown of the glycosides. HCN is readily absorbed into blood and is distributed to all tissues of the body.

Heliophytes: The plants that thrive in area where intensity of light is more are called heliophytes

Hemic organic soil materials: Intermediate in degree of decomposition between the less decomposed fibric and the more decomposed sapric materials.

Herbicidal Efficiency Index (HEI): It is calculated as follows:

$$\{(Y_T - Y_C/Y_T)\ x100/(WDM_T/WDM_C)\ x\ 100\}$$

Where, Y_T is crop yield from the treated plot; Y_C is crop yield from the weedy check; WDM_T is weed dry matter in the treated plot; and WDM_C is weed dry matter in the weedy check plot.

Herbicide Resistant Crops (HRC_S): HRC_S are transgenic crops are designed to tolerate specific broad-spectrum herbicides, which kill the surrounding flora, but leave the cultivated crop intact. These transgenic crops contain genes that enable them to degrade the active ingredient in an herbicide, rendering it harmless. Currently, herbicide resistant crops are common for soybean, maize, repesees and cotton-Roundup Ready (active agent: glyphosate) and Liberty Link (active agent: glufosinate).

Herbicide Rotation: It is a practice of systematic use of herbicides having different mode of action, but same level of activity, selectivity and persistence and targeting similar spectrum/group of weeds in sequence in the same field to prevent insurgence of herbicide resistant weeds and shift in weed flora and to control weeds/resistant weeds effectively. An effective herbicide rotation provides greater weed control efficiency for a long period all through the crop growing cycle and higher cumulative benefit: cost ratio.

Herbicide Therapeutics: Herbicide molecules possessing certain plant disease suppression effects.

Herbicide-Resistant Weeds: Weeds which were earlier susceptible to an herbicide, but gradually developed resistance to that herbicide in course of time as a consequence of frequent and continued use of the same herbicide year after year to control them are herbicide-resistant weeds. This is biochemical mimicry.

Herbicide-Tolerant Weed: A weed which is normally not controlled while others are readily controlled by a herbicide is a tolerant one. This tolerance to a certain herbicide as observed in crop plants is a genetically inherited trait and not an acquired one on continuous exposure to herbicide. Many perennial weeds (*e.g. Cynodon dactylon* and *Cyperus* sp.) are normally tolerant to the selective pre-planting, pre-emergence or post-emergence herbicides at their usual field rate. Certain annual weeds, such as *Digera arvensis* and *Digitarias anguinalis* are tolerant to pendimethalin; and *Acrachnera cemosa* is tolerant to atrazine.

Herbigation: Application of herbicide along with irrigation water is herbigation.

Hidden Hunger: It is a term used to describe a plant that shows no obvious symptoms yet the nutrient content is not sufficient to give profitable yield.

High Analysis Fertilizer: Having more than 25 per cent of the primary nutrient content *e.g.*, Urea, anhydrous ammonia.

Histosols: Soils formed from materials high in organic matter, histisols with essentially no clay must have at 20 per cent organic matter (12 per cent C) by weight (about 78 per cent by volume). This proportion increase to 30 per cent (85 per cent by volume) in soils with at least 50 per cent clay.

Homeostasis: The tendency towards a relatively stable equilibrium between interdependent elements, especially as maintained by physiological processes.

Horizonation: The Development of all the horizons in soil is Horizononation

Humic Acid: A term of varied usage but usually referring to a mixture of indefinite composition of dark coloured organic substance precipitated upon acidification of a dilute alkali extract of soil.

Humification: Process of decomposition of organic matter leading to the formation of humus.

Humus: The well decomposed, more or less stable part of the organic matter of the soil.

Hydrogen Bonding: H-bonding is a bond between two highly electronegative atoms, *e.g.* oxygen through hydrogen (H) atom and a good example is water (H_2O). The same kind of arrangement is operative between oxygen atoms of carboxylic acid groups (-COOH or –COO after ionization) of the herbicides with hydrogen atoms of the amino groups ($-NH_2$ or NH_3^+ after ionization) or of the alcohol groups (-OH) of the organic matter.

Hydrologic Cycle: The fate of water from the time of precipitation until the water has been returned to the atmosphere by evaporation and is again ready to be precipitated.

Hydrological Drought: Meteorological drought, when prolonged results in hydrological drought with depletion of surface water and consequent drying of reservoirs, tanks *etc.* This is based on water balance and how it affects irrigation as a whole for bringing crops to maturity.

Hydrophilic: Colloidal material which attracts or love water.

Hydrophobic: Colloidal materials which have the properties of water repellent.

Hydroseeding: Planting seed in a water mixture by pumping through a nozzle that sprays the mixture onto a seedbed. The water mixture may also contain addends such as fertilizer and mulches.

Hygroscopic Coefficient: The hygroscopic coefficient is the maximum amount of hygroscopic water absorbed by 100 g of dry soil under standard conditions of humidity (50 per cent relative humidity) and temperature (15°C). This tension is equal to a force of 31 atmospheres

Hypogeal: The germination process in which no part of the cotyledon comes out of the soil is hypogeal germination. It occurs in monocot grass species.

Hysteresis: It is the lag in one of the two associated processes or phenomena during reversion in the soil moisture absorption and desorption process *i.e.*, difference in moisture content at a given suction.

Ideotype: Ideotype is a biological model which is expected to perform or behave in a particular manner within a defined environment.

Igneous rocks: The lava on the surface cools down and gets solidified as rocks called igneous rocks. Granite and basalt are examples of such rocks. Igneous rocks are also called Primary or Mother rocks because all other rocks are directly or indirectly formed from them.

Illuviation: The process of accumulation of gravel deposits, sand, silt or clay, at places in rivers, lakes, or estuaries were the flow is checked.

Imbibition: Certain substances if placed in a liquid absorb it and swell up. These substances are called as imbibants and the phenomenon is imbibition's, certain force of attraction is existing between imbibants and the involved substance.

Immobilization: It is the conversion of inorganic compounds to organic compounds by micro-organisms or plants, by which it is prevented from being accessible to plants.

Importance Value Index (IVI): The importance value index (IVI) of a plant species is calculated to appraise the ecological importance of a species in relation to community structure of a community. It is the sum-total of relative density, relative dominance and relative frequency. IVI thus calculated is a value out of 300 for each species of composite stand. IVI= [Relative frequency + relative density +relative dominance)].

Income Equivalent Ratio (IER): The ratio of the area needed under sole cropping to produce the same gross income as one hectare of intercropping at the same management level. The IER is the conversion of LER into economic terms.

Incomplete fertilizer: Containing any two primary nutrients.

Infiltration Rate: The rate of water entry into the soil expressed as a depth of water per unit of time and expressed in inches per hour or feet per day. The infiltration rate changes with time during irrigation.

Inflorescence: The complete flower head of a plant including stems, stalks, bracts, and flowers.

Inhibitor: A chemical substance that acts to prevent a process from occurring.

Inoculate: To treat, usually seeds, with microorganisms to create a favourable response. Most often refers to the treatment of legume seeds with *Rhizobium* or *Brady rhizobium* to stimulate dinitrogen fixation.

Inverse nitrogen yield concept: According to Wilcox, the yield of the crop is inversely proportional to its nitrogen content, *e.g.* $Y= K^{-n}$, where Y is the yield, n is the N (per cent) in crop, and K is the constant (318 lb/acre).

Invisible drought: This can occur even when there is frequent rain in an area. When rainfall is inadequate to meet the evapotranspiration losses, the result is borderline water deficiency in soil resulting in less than optimum yield. This occurs usually in humid region.

Ion mobility in soils: Anions (negatively charged ions) usually leach more readily than cations because they are not attracted to the predominantly negative charge of soil colloids.

For example, NO_3^-, due to its negative charge and relatively large ionic radius, is not readily retained in the soil and is easily lost from soils by leaching.

Irrigation Efficiency: It is the ratio expressed in percentage of water stored in the root zone depth of soil to the water delivered in the field from farm supply source.

Irrigation Frequency: This refers to the number of days between irrigations during periods without rainfall. It depends on the consumptive use rate of a crop and on the amount of available moisture in the crop root zone.

Irrigation Interval: Time between the start of successive field irrigation application and the same field, days.

Irrigation Requirement: It is the part of total water requirement of a crop, exclusive of effective rainfall and soil moisture stored on the root zone or that contributed from the shallow groundwater table. Irrigation requirement of a crop refers to the water requirement of crops, exclusive of effective rainfall and contribution from soil profile.

IR = WR – (ER + S) where, IR is irrigation requirement, WR is water requirement, ER is effective rainfall, and S is soil moisture stored in the root zone.

Irrigation Scheduling: Amount and time of application of irrigation according to crop need and water availability.

Isobars: Lines of equal pressure.

Isobutyledenediurea (IBDU): A reaction product of urea and isubutyledene containing about 30 per cent N in the fertilizer grade. A slow release N-fertilizer.

Isohels: Imaginary lines joining points of equal sunshine hours.

Isohyet: A line drawn on a map connecting points with equal rainfall.

Isolated Field: A field where the rate of soil flow is zero at the windward edge of the field due to the presence of a stable border. An isolated field is not protected by barriers and is exposed to open wind velocities. The Wind Erosion Equation applies to conditions on an isolated field.

Isoline: A line on a map or chart along which there is a constant value of a variable such as wind velocity or climatic erosivity.

Isopleths: A line of equal line of a given quantity.

Isostere: Imaginary lines connecting points of constant moisture content.

Isotach: A line of equal wind speed

Isotherm: Imaginary lines connecting points of equal temperature.

IW:CPE ratio: Based on the climatological approach, scheduling of irrigation water is done and the depth of irrigation water is fixed for different crops.

$$= \frac{\text{Irrigation water (mm or cm)}}{\text{Cumulative pan evaporation (mm or cm)}}$$

Jalshakti: A chemical which when applied (mixed) in soil, improves the aeration, infiltration and water holding capacity of the soil.

Jhum cultivation: The slash and burn type of shifting cultivation in the hill tracks of Bangladesh and north eastern part of India.

J-shaped/Geometric Curve: It is an intermediate form/shape of the population growth curve observed in some plants and deviation from the sigmoid curve of population growth. The growth form/curve varies mainly on the basis of nature of plant species and environment.

Kaolinite: An aluminosilicate mineral of the 1:1 ratio crystal lattice group, that is consisting of one silicon tetrahedral layer and one aluminium octahedral layer of about 0.7nm thickness.

K-Selected Organisms/K-Selections: The organisms which are more stable, long-lived (*e.g.* shrubs, trees), have prolonged vegetative stage, adapted for persistence and reproduction in stable environments, a small portion o their biomass is allocated to reproduction, and are less likely to suffer from high mortality of immature individuals are K-selected ones. Large-seeded annuals and many perennial weeds have the K-strategy. "K" is the highest population density that can be maintained in a real environment (at carrying capacity) when population is in equilibrium.

Kurtosis: It is a measure of 'peakedness' of the probability distribution of a real valued random variable. The departure of peakedness of a symmetrical frequency distribution from the normal by excess (lepto kurtosis) or deficiency (platy-kurtosis) in its shoulders as opposed to tail and centre.

Kyoto protocol: An international agreement on climate change that sets a target for signatory countries' collective emission reductions and mechanism for doing so. The agreement was reached in 1997 in Kyoto, Japan and came into effect in February 2005.

L value: It is a measure of the isotopically exchangeable fraction of phosphate in soil and is used to determine the quantity of soil phosphate in labile form.

Land equivalent coefficient (LEC): The index was proposed by Adetiloye and Ezedinma (1983). The theoretical maximum value of LEC is unity when mixtures become ideally complementary.

$$LEC = LER_a \times LER_b = 0.5 \times 0.5 = 0.25. \ LEC = LER_a \times LER_b = 1 \times 1 = 1.$$

Land equivalent ratio (LER): It denotes the relative land area under sole crop required to produce the same yield as obtained under a mixed or an intercropping system at the same management level.

$$LER = (Yab/Yaa) + (Yba/Ybb)$$

Where, Yab is the yield of species of 'a' in association with species 'b', and Yba is the yield of species 'b' in association with species 'a', Yaa and Ybb represent the pure stand yields of species 'a' and 'b' respectively.

Land: The part of surface of lithosphere which is not usually covered with water forms the total natural and cultural environment with which production takes place.

Lapse Rate: The rate at which temperature decreases in a rising and expanding air parcel, *i.e.* temperature decreases with increase in altitude (6.5^{0}C/km).

Latent Heat: The heat released or absorbing per unit mass by a system during a change of phase. In meteorology, at 0^{0}C the latent heat of vaporisation, fusion and sublimation of water are about 540, 80 and 680 cal/g, respectively.

Laterite Soil: Soil formed by *in situ* by the leeching out of the bases from the parent rock under monsoon conditions of alternate dry and wet seasons. These are characterized by poor fertility and deficient in N, P, K and Ca.

Laterization: Process of leeching out of completely of calcium at or near the surface of the soil in heavy rainfall areas under tropical conditions, due to intense weathering takes place to a great depth. Iron and aluminium concentrate at or near the surface of the soil.

Lathyrism: Paralytic condition of animals caused by consumption of forage or seed of certain plants of genus *Lathyrus*.

Law of Average Product: Average product is the total output divided by the total input.

Law of Constant Returns: Constant productivity holds true if all units of the variable factor which are applied to the fixed factor result in equal additions to total output.

Law of Diminishing Returns: Diminishing productivity of the variable factor exists when each additional unit of the input acts less to output than the previous unit.

Law of Minimum: It states that when a character such as yield is controlled by number of separate factors, the rate of the process is limited by the factor which is present in relative minimum amount.

Lay-by application: It is the application of herbicide with or after the last cultivation in crops, *e.g.* after ridging in sugarcane and cotton.

Layout: The placement of experimental treatment of the experimental site where it may be over space, time or type of material.

Leaching Index: The leaching index (LI) is a simple tool for estimating the potential for percolation below the root zone for a particular site (soil, precipitation, and climate), and thus the potential for movement of soluble materials below the root zone.

Leaf Colour Chart (LCC): It is a diagnostic tool for monitoring the relative greenness of a rice leaf as an indicator of the plant N status. The LCC is usually a plastic ruler shaped strip containing four or more panel that range in colour from yellowish green to dark green.

Leghaemoglobin: A iron rich compound found n the root nodules of leguminous plant.

Legumes: Legumes, through the symbiotic relationship they have with various soil organisms (primarily *Rhizobium* spp.), can be a valuable source of nitrogen for a succeeding crop. The soil organisms, which invade the root hairs and receive energy from the plant, fix nitrogen from the atmosphere that is later available for the growing legume crop and to succeeding crops.

The proper species of *Rhizobium* must be present in the soil or applied with the seed to ensure that fixation will occur at rates which supply the N requirements of the legume crop and make N available to succeeding crops.

Leibig's law of minimum: The growth and reproduction of an organism are determined by the nutrient substance that is available in minimum quantity, the limiting factor. In general it states that when a character such as yield is controlled by number of separate factors, the rate of the process is limited by the factor which is present in minimum amount.

Ley farming: A rotation of arable crops requiring annual cultivation and artificial pastures occupying field for two years or longer.

Liebig's Law: The growth and reproduction of an organism is dependent on the nutrient substance that is available in minimum quantity.

Lifesaving Irrigation: The irrigation given to a dryland or rainfed crop to save it from a total failure due to long dry spell or severe soil moisture stress, the source of irrigation generally being harvested water.

Light compensation point: The light intensity at which photosynthesis and the respiration rates in a leaf or other chlorophyllous organ are equal.

Light Soil: A soil that is easy to work implements in o r easy to cultivate as it lacks cohesiveness, such as sandy and sandy loam soils.

Lignin: An organic chemical which occurs in the cells, especially wood cells and make substance resistant to decomposition.

Liming: Liming is a critical management practice for maintaining soil pH at optimal levels for growth of plants. Over-liming can induce micro-nutrient deficiencies by increasing pH above the optimum range. Lime has its maximum effect in a period of 1 to 3 years, while suspension lime, burned lime, and hydrated lime have their maximum effect in 3 to 6 months.

Example: Hydrated ($Ca(OH)_2$), or burned (CaO) lime, which will neutralize soil acidity quickly.

Lint Index (LI): The weight of lint per seed cotton or 100 seeds

L_I = weight of 100 seeds (seed index) x ginning percentage (GP)/100 – GP

Liquid Limit: It is the state at which the film of water is thick enough to reduce cohesion between soil particles.

Lister: An implement for furrowing land, often a planting attachment.

Lithophytes: Lithophytes are plants that grow in or on rocks.

Loam: Soil material that contains 7 – 27 per cent clay, 28 – 50 per cent silt and less than 52 per cent sand.

Loamy: A broad grouping of soil texture classes includes sandy loam, clay loams, loamy silt and silt loam texture.

Local Control: Making use of greater homogeneity of adjacent areas in a field, the entire experimental units are divided into compact blocks which is know as local control. It helps in reducing the experimental error, thereby increasing the precision of the experiment.

Longevity: It is the life span of weed seeds in soil which is influenced by variable and long dormancy periods of seeds, presence of seeds at various depths of soils experiencing various edaphic conditions, and variable and longer viability of seeds attributed to hard seed coats or seeds resisting decay decomposition. There is a great variation in the longevity of seeds in soil across weed species and environmental conditions.

Lopping: Pruning or cutting away aerial part of shrubs or trees.

Low Analysis Fertilizer: Having less than 25 per cent of the primary nutrient content *e.g.,* SSP (16 per cent P_2O_5).

Luxury Consumption: The absorption of nitrogen or potash from the soil by a crop in excess of crop needs.

Main Effect: An estimate of the effect of an experimental variable or treatment measured independently of other which may form part of the experiment. Thus, in a factorial experiment involving three factors A, B and C each at two levels (applied and not applied),

The main effect due to A would be the average o the four effects where A was applied and B and C were applied at each of the two levels, less the average of the four effects where A was not applied.

Manure: The excreta of animals with straw or other materials, used as the absorbent. The decomposed manure is called as farmyard manure or farm manure or barnyard manure. The average composition of well rotted farmyard manure is 0.5 per cent N, 0.3 per cent P_2O_5 and 0.5 per cent K_2O.

Marling: A friable earthy deposit consisting of clay and calcium carbonate, used especially as a fertilizer for soils deficient in lime.

Marsh: A periodically wet or continually flooded area with the surface not deeply submerged. Covered dominantly with sedges, bushes or other hydrophytic plants.

Mean: It is the ratio of sum of all values o a series to that of the number of values in that series. The mean is usually denoted by μ.

Measures of Central Tendency: Statistical measures used to represent the centre of a data set are called measures of central tendency. The commonly used measures of central tendency are: Mean (arithmetic, geometric or harmonic), median and mode.

Median: The median is the value, which is located in the middle of a series when the observations are arranged in order of magnitude. It divides the series into two equal halves, half the number of observations lying above it and half below. It is not affected by extreme observations.

Mellow Soil: A soil that is easily worked or penetrated.

Mesopause: The upper boundary of Mesosphere is called the "Mesopause".

Mesosphere: The mesosphere is a layer of Earth's atmosphere. The mesosphere is directly above the stratosphere and below the thermosphere. It extends from about 50 to 85 km (31 to 53 miles) above our planet. Temperature decreases with height throughout the mesosphere.

Metamorphic Rock: When igneous or sedimentary rocks are subjected to extreme heat and pressure, they undergo a complete change in their form and character.

Meteorological Drought: It is defined as a condition, where the annual precipitation is less than the normal over an area for prolonged period.

Methaemoglobinaemia: A condition associated with nitrate toxicity as a result of consuming nitrate rich waters. Nitrate transforms ordinary oxy-haemoglobin into inactive methaemoglobin and infants suffering from this condition referred to as "blue babies" due to development of greyish-blue skin, and the problem as "**blue baby syndrome**".

Methane: A gas produced in very highly reduced (Eh <244 mv) paddy soils. One of the gases blamed for contributing to global warming through the greenhouse effect.

Micelle: The minute silicate clay colloid particles, ordinarily carrying negative charges is called micelles.

Microbial Degradation: It is breakdown or detoxification and inactivation of the soil applied herbicides by soil micro-organisms.

Microemulsion: Microemulsions are clear, thermodynamically stable, isotropic liquid mixtures of oil, water and surfactant, frequently in combination with a co surfactant.

Micronaire: This is the size of an individual cotton fibre taken in cross section. Micronaire is an indicator of air permeability.

Milliequivalent per Litre (me/l): A milliequivalent of an ion or a compound in one litre of water. Milliequivalent = equivalent weight/1000.

Mimicry: It is a pretending to be what an individual is really not. In natural system, some plants pretend in various forms to maintain their survival and continued existence, *e.g.* pitcher plant (venus fly trap), sundew *etc.* In crop fields too, some weeds mimic with crop plants in various forms to ascertain their existence and perpetuation with crops.

Mineral soil: A soil composed mainly of, and having its properties determined by, mineral matter, with less than 20 percent organic matter. Compare Organic soil.

Mineralization: Mineralization in soil science is the decomposition of the chemical compounds in organic matter, by which the nutrients in those compounds are released in soluble inorganic forms that may be available to plants.

The conversion of an element from an organic form to an inorganic state as a result of microbial activity.

Minimum Tillage: Minimum tillage is aimed at reducing tillage to the minimum necessary for ensuring a good seedbed, rapid germination, a satisfactory stand and favourable growing conditions.

Mistscherlich's Law: It assumes that a plant produces its maximum yield if all conditions are ideal if any essential factor is absent, there is corresponding reduction in the yield.

$$dy/dx = (A-Y) \times C$$

where Y is the yield, X is the amount of factor present, A is the maximum yield and c is the constant.

Mixed Farming: Defined as a system of farming on a particular farm which includes crop production, raising livestock, poultry, fisheries, bee keeping *etc.* To sustain and satisfy as many needs of the farmer as possible.

Mixed Fertilizer: Individual or straight fertilizer materials are blended together physically to permit application in the field in one operation. Such fertilizers supply two or three major nutrients in a definite proportion of grade, *e.g.* Nitro phosphate with potash 15:15:15 of NPK.

Mode: The mode is that value of the variate which occurs most often or with the maximum frequency.

Moisture Deficit Index (MDI): This is a measure to estimate the dryness of a region and is expressed by the formula, $MDI = [(P-PE)/PE] \times 100$; where, P is rainfall (mm), and PE is the potential evapotranspiration (mm).

Moisture Equivalent: The weight percentage of water retained by a previously saturated sample of soil 1 cm in thickness after it has been subjected to a centrifugal force of 1000 times gravity for 30 minutes.

Moisture Regimes: The different levels of the available soil moisture (per cent), a testing factor or unit inputs used in conducting irrigation experiments, such as ASM at 100, 80, 60, 40 per cent, *etc.*

Molal Solution: A molal solution is a solution that contains 1 molecular weight of solute in a kilogram of solvent. It is the strength or concentration of a solution, especially the amount of dissolved substance in a given volume of solvent. It is a concentration of a solution expressed in moles or molality (m).

Molar Solution: A one molar solution is a solution in which 1 mole of a compound is dissolved in a total volume of one litre.

Molasses: Molasses is a thick syrup that people use as a sweetener. It is a by product of the sugar-making process, and it comes from crushed sugar cane or sugar beets.

Molecule: The smallest unit of a substance that can independently exist and retain its properties.

Monetary Index: A single productivity index which incorporates monetary value of all crops in a system.

Monitory Advantage Index (MAI): An index was formed by taking into consideration the relative money value of the produce.

MAI = LER – 1/LER x value of combined intercrops

Monocropping and monoculture: A cropping system in which the same crop is grown year after year in the same field.

Monoecious: Majority of weeds are monoecious which means male and female flowers are in same plant but in different locations, *i.e.* androecium and gynoecium/pistil are not in same flower and as a result, no complete autogamy exits.

Mulch: Any material such as straw, sawdust, leaves, plastic film, loose soil, or similar material that is spread or formed upon the surface of the soil to protect the soil and/or plant roots from the effects of raindrops, soil crusting, freezing, evaporation, *etc.*

Mulch tillage: Managing the amount, orientation, and distribution of crop and other plant residue on the soil surface year-round, while growing crops where the entire field surface is tilled prior to planting.

Multiple Correlation: Multiple correlation coefficient measures the collective association of several variables with one another. It is the correlation coefficient between the observed value of the dependent variable and its predicted value in a multiple linear regression.

Multiple Cropping Index: It measures the sum of areas planted to different crops and harvested in a single year divided by total cultivated area times 100.

Multiple Regression: The regression of a dependent variable on more than one 'independent' or 'predictor' or 'explanatory' variable.

Multi-storey Cropping: Cultivation of two or more than two crops of different heights simultaneously on a certain piece of land in any certain period *e.g.* Sugarcane + Mustard + Onion/Potato.

Municipal Solid Waste: Municipal solid waste can include anything from shredded cardboard and newspapers to grass clippings and leaves. These materials vary widely in nutrient content (from 0.1 per cent N up to 3.4 per cent N), moisture content and C:N ratio. The important characteristic is C:N ratio. Materials with a high C:N ratio, such as corrugated cardboard, take longer to decompose and require lots of energy by soil microorganisms to decompose them.

Musk Soil: Fairly well decomposed organic soil material relatively high in mineral content, dark in colour and accumulated under conditions of imperfect drainage.

N:S Ratio: The ratio between organic N and S in plant material or soil. In plants this is generally 10 - 15. In soil organic matter, it is generally 7-10.

Net Area Sown: When cropped more than once in a any year, it is area is added to the net area sown to arrive at the total area sown during the year.

Net Cropped Area: Area sown with crops but is counted only once

Net Irrigated Area: The net irrigated area is the actual land area on which irrigation was used for growing crops for as many times as many in an agricultural year.

Net Irrigation Requirement (NIR): The depth of irrigation water required to bring the soil water level to field capacity in the effective root zone depth. The net irrigation requirement excludes precipitation, carry over soil moisture or groundwater contribution or other gains in the soil moisture.

Net Photosynthesis: When the photosynthesis is measured with correction for respiration.

Net productivity: The arithmetic difference between calories produced in photosynthesis and calories lost in respiration.

Net Return: It gives the actual profit obtained by the farmer. This is worked by subtracting the total cost of cultivation from the gross returns.

Neutron Moisture Meter (Depth probe): Neutron source and detector in a cylindrical container that can be lowered down a tube in soil or other material and calibrated to indicate moisture content.

New Rice for Africa: It is a cultivar proof of interspecific hybrid rice created by crossing *O. glaberrima* and *O. sativa* developed by the Africa rice centre (Africa rice to improve the yield of African rice cultivars.)

Niche: Niche refers to overall place/position/status of a species in a community, which may include its place in the space, time and function of that community. Niche differentiation thus means making association or groups of species on the basis of their places in a community.

Nick: In hybrid seed production, the parents are said to nick when they produce higher yield as a result of synchronized flowering.

Nipping: It is a special cultivation practice of plucking the apical buds of the crop at about 30-40 days after sowing. Nipping stops the apical growth and promotes the lateral branching, thus the plants become more vigorous and produce more flowers and pods and yield per plant is increased.

Nitrate – Nitrogen: Presence of nitrogen in forms of nitrate, denoted (NO_3–N). It occurs in soil medium as the ammonia is converted into nitrate due to microbial activity (*Nitrosomonas* and *Nitrobacter*). Nitrate fertilizers are carriers of nitrate nitrogen. Most of the plant species use nitrate nitrogen.

Nitrification Inhibitor: The chemicals used to control nitrogen transformations in the soils. Due to their toxic effects on nitrifying bacteria, the temporarily inhibit the production of nitrate, *e.g.* N – saver, nitrapyrin, potassium azide *etc.* application rate 0.32 – 0.6 kg of active ingredient.

Nitrification: It is a two step process which the ammonia or ammonium is converted to nitrite and finally to nitrate and these two steps are carried out by autotrophic micro-organisms *Nitrosomonas* and *Nitrobacter*, respectively.

Nitrogen Assimilation: The incorporation of nitrogen compounds into cell substance by living organisms.

Nitrogen Cycle: Nitrogen is subject to more transformations than any other essential element. These cumulative gains, losses, and changes are collectively termed the nitrogen cycle. The ultimate source of nitrogen is N_2 gas, which comprises approximately 78 percent of the earth's atmosphere. Inert N_2 gas, however, is unavailable to plants and must be transformed by biological or industrial processes into forms which are plant-available.

Nitrogen Fixation: Nitrogen fixation is the process whereby inert N_2 gas in the atmosphere is transformed into forms that are plant-available, including NH_4^+ or NO^-. Fixation can take place by biological or by non-biological processes.

Nobilization: Nobilization refers to the breeding theory of development and utilization of wild germplasm. *Saccharum spontaneum* is the main donor of resistance and adaptive genes in the nobilization breeding process.

Nodule: A structure developed on the roots of most legumes and a few other plants in response to the stimulus of root-nodule bacteria, usually genus *Rhizobium.*

Non-Symbiotic Nitrogen Fixation: This is a N_2 fixation process that is performed by free-living bacteria and blue-green algae in the soil. The amount of N_2 fixed by these organisms is much lower than that fixed by symbiotic N_2 fixation.

Normal Distribution: The normal distribution, also called the Gaussian distribution, is the most important distribution in statistics. Most of the theories developed assume that the random variable is normally distributed. A random variable X is said to have a normal distribution with mean μ and variance σ^2 if its probability density function is given by:

$$f(x)dx = \frac{1}{\sigma\sqrt{2\pi}}\exp\left\{-\frac{1}{2}\frac{(x-\mu)^2}{\sigma^2}\right\}, \quad -\infty < x,\ \mu < \infty,\ \sigma > 0$$

Mean μ and variance σ^2 are the parameters of the normal distribution. This distribution is symmetrical about mean and is mesokurtic. Mean, median and mode coincide. The standard normal distribution is the normal distribution with mean zero and variance one.

Normal ploughing: Ploughing upto a depth of 15 cm.

Normal Rainfall: Rainfall value between +19 per cent to -19 per cent of the normal value of season rainfall.

Normalized Difference Vegetation Index (NDVI): Acronym for normalized difference vegetation index, which provides a crude estimation of vegetation health and means of monitoring the changes in vegetation over time. The possible values range of values are between -1 and 1, but the typical range between about 0.01 (NIR less than VIS for not very green area) to 0.6 (for a very green area).

Null Hypothesis: The Statistical hypothesis under sample study is called null hypothesis. It is usually that the observations are the result purely of chance. It is usually denoted by H_0. A null hypothesis is a hypothesis the possibility of rejection of which is of interest to the researcher.

Nurse Crop: A companion crop which nurses the main crop by way nitrogen fixation and slash or adding the organic matter into the soil, *e.g.* cowpea intercropped with cereals or new plantation of fruit trees.

Nursery: Series of plant progenies, introductions *etc.* grown chiefly for individual plant testing and examination in a plant breeding programme. In the nursery-bed, seedlings are raised for transplanting to the main field.

Nutrient Balance: It is the difference between nutrient additions to soil and removal.

Nutrient management: Nutrient management is defined as managing the rate, timing, form, and method of nutrient application to ensure adequate soil fertility for plant production and to minimize the potential for environmental degradation, particularly air, soil, and water quality impairment. Nutrient management is the implementation of management techniques that permit efficient crop production while protecting natural resource quality. Nutrients are considered any element or compound that are essential for plant growth, particularly the elements nitrogen (N), phosphorus (P), and potassium (K). Nutrient sources can be any material, such as fertilizers, animal manures, biosolids, and irrigation water that contain essential plant nutrients.

Nutrition: The sum of the process by which an animal or plant absorbs or take in and utilizes food substances.

O – horizon: Organic horizon of organic soils.

Objectionable Weed: A problem weed; whose seed once mixed with crop seed is extremely difficult to separate.

Off-Season Nursery: Growing the breeding materials during the off-season (a season in which the crop is not grown in the concerned locality) in a location suitable for the purpose generally to advance the generation.

Off-Type: Plants or seeds deviating significantly from the characteristics of a variety as described by the breeder in any observable respect.

One-Tailed Test: A test of hypothesis in which the alternative hypothesis is such that it provides one sided alternative. If it provides left sided alternative, then it is called *'left tailed test'.* If the alternative is right sided, then it is known as *'right tailed test'.* For instance, testing H_0: Effect of treatment i = Effect of treatment j versus H_0: Effect of treatment i< Effect of treatment j leads to a left tailed test.

Open Pan Evaporation (Eo): It is the value of evaporation from an open pan (standard USWB Class 'A' pan) evaporimeter. Water needs of crops can be well predicted from the values of open pan evaporation in a given area and in a specified period.

Opportunity Cropping: The practice of placing primary emphasis on the use of stored soil moisture while determining whether or not to establish a crop.

Organic Matter: It includes plant and animal residue of various stages of decomposition in soil, cell and tissue of soil organisms, substance synthesized by soil microbial activity. Plants leaves, stem, straw, husk, root, stubbles or any remnant of crops or plant species and organic manures including farm yard manure – are the cheap source of organic matter.

Organic Nitrogen: Nitrogen in the form of organic compounds, *e.g.* amino acids, amino sugars, purine and pyrimidine bases.

Organic Phosphorus: Phosphorus present as a constituent of an organic compound or a group of organic compounds, *e.g.* glycerophosphoric acid, inositol phosphoric acid, cytidylic acid.

Organic soil: Soil containing organic matter in sufficient quantities to dominant the soil characteristics frequently all soils containing 20 per cent are more organic matter by weight are arbitrarily designed as organic soils.

A soil that contains a high percentage (greater than 20 per cent) of organic matter throughout the solum. Compare Mineral soil.

Oriented Tillage: Tillage operations which are oriented in specific parts or directions with respect to the sun, prevailing winds, previous tillage actions or field base lines.

Osmotic Potential: The amount of work that must be done per unit quantity of water in order to transport reversibly and isothermally an infinitesimal quantity of water from a pure water at specified elevation and atmospheric pressure to a pool containing a solution identical in composition with soil water but in all other respects identical to reference pool.

Osmotic Pressure: The pressure developed due to unequal concentrations of salts separated by a semi permeable membrane or plant cell wall. Water will move from the side of lower salt concentration or higher free energy through the membrane to the side of high salt concentration or low energy. This water exerts additional pressure termed as osmotic pressure (numerically equal and opposite to solute potential).

Oven-Dry Weight: The weight of a substance after it has been dried in an oven at 105°C, to equilibrium.

Oxidation: It is an energy yielding process usually resulting from loss of hydrogen or addition of oxygen to a compound. The electron acceptor is called as oxidant.

Oxygen Diffusion Rate (ODR): It is one of the criteria generally used to determine the oxygen concentration in the soil pore spaces.

Paira Cropping: Paira cropping is sowing of second crop before harvesting of the main crops

Paired Row Planting: In this technique two adjacent rows of the base crops are paired, reducing the inter-row space in the pair narrow enough to create some inter-space between pairs between pairs of base crops rows but wide enough to minimise undue competition among plants.

Pan Coefficient (Kp): Ratio between reference evapo transpiration ET_0 and water loss by evaporation from an open water surface of a pan (Epan); expressed as $Kp = ET_0/Epan$.

Pan Evaporation: It is the value of evaporation from an open pan evaporimeter, usually from standard USWB, class 'A' pan evaporimeter during a day or specified period at a place.

Parameter: Refers to the quantity that represents a particular characteristic of the population. Customarily, the parameters are represented by Greek letters. Population mean and population variance are some examples of a parameter.

Parthenocarpy: The development of fruit without fertilization are phenomenon of seedless fruit production.

It is a mechanism in which seeds are produced without pollination and fertilization. It operates in some plats, *e.g. Taraxacum officinale* (common dandelion).

Particle Density: Particle density is defined as the ratio of mass of solids to the volume of the solids alone and is expressed as g/cm^3.

Partition Coefficient (K_D): It is the ratio of herbicide bound/adsorbed to soil colloids to herbicide left in soil solution. Smaller the K_D value, greater is the concentration of herbicide in soil solution and *vice-versa.* Thus, the partition coefficient is inversely related with the amount of herbicide present in soil solution and it gives an indication of adsorption of herbicide in soil. (K_D=Herbicide held in soil colloids/Herbicide in soil solution).

Pedigree Breeding: System of breeding in which individual plants are selected in the segregation generations from a cross on the basis of their desirability judged individually and on the basis of pedigree record.

Pegging: Groundnut budding ovary is called a "peg." The peg enlarges and grows down and away from the plant forming a small stem which extends to the soil. The peanut embryo is in the tip of the peg, which penetrates the soil is called pegging.

Penetrometer: A *penetrometer* is a device to test the strength of a material.

Perennial: The plant that dies back seasonally and produce new growth again from a perennial part.

Permanent Drought: This is characteristic of the desert climate where sparse vegetation growing is adapted to drought and agriculture is possible only by irrigation during entire season.

Permanent Wilting Point: Permanent wilting point otherwise called as soil-water content at which most plants cannot obtain sufficient water to prevent permanent tissue damage. The lower limit to the available water capacity has been reached for a given plant when it has so exhausted the soil moisture around its roots as to have irrecoverable tissue damage, thus yield and biomass are severely and permanently affected. The water content in the soil is then said to be the permanent wilting percentage for the plant concerned.

Permeability: The ease with which water, air, or plant roots penetrate or pass through a soil horizon. Permeability is the measure of the resistance of a material against the formation of a magnetic field, otherwise known as distributed inductance in transmission line theory.

Permeameter: A device for measuring the permeability of soils or other materials.

Persistence: Ability to withstand extremes of climatic, edaphic, and biotic stresses in nature. The maintenance of the species is achieved by vegetative spread or by regeneration from self-sown seed.

Pesticides: Chemical compounds that are used to kill pests, including insects, rodents, fungi and unwanted plants (weeds).

pF: It is the logarithm of height in cm of water column that represents total stress with which water is held by a soil.

pH: Refers to the potential of hydrogen, defined as the negative logarithm of the hydrogen ion concentration of a solution. $pH = -\log_{10}(H^+)$.

Phenology: Phenology is concerned with certain plant events (*e.g.* growth, flowering, fruiting *etc.*) occurring at a particular time/season.

Phenotype: Physical expression or outward appearance of a genotype which is subject to change with the environment.

Phosphorus Index (PI): The phosphorus index was conceived as a field-based method for determining the relative potential for P movement from a site.

Photo-oxidation: When the light intensity exceeds a limit, leaves consume oxygen and use it in the oxidation of certain cell constituents. It is a chlorophyll catalysed process and more chlorophyll molecules become excited than possibly be utilized causing harmful side effects. It results in bleaching of chlorophyll and inactivation of important enzymes.

Photoperiodism: Photoperiodism is the physiological reaction of organisms to the length of night or a dark period.

Phyllochron: It is the time taken by rice plant to form a new tiller with leaf. If all the conditions are favourable one phyllochron is completed in 5 to 7 days.

Physiological Drought: Non-availability of water to the plant due to unfavourable physiological conditions, such as water logging, soil salinity, low temperature *etc.*

Piezometer: Instrument used for measuring depth of water table.

Pisciculture: The breeding, rearing, and transplantation of fish by artificial means is called pisciculture, in other words, fish farming. It is the principal form of aquaculture, while other methods may fall under mariculture. It involves raising fish commercially in tanks or enclosures, usually for food.

Planking: One of the tillage operations carried out before sowing the crop for smoothening, micro levelling and cover the seeds after sowing.

Plant Analysis: Plant analysis is the laboratory determination of elemental composition of a sample of plant tissue. This technique is most commonly used to diagnose nutritional problems related to soil fertility or to monitor the effectiveness of fertilizer practices on growing crops.

Plant Available Nitrogen: This plant available nitrogen is the actual amount of N in the biosolids that is available to crops during a specified period.

Plant Ecology: The branch of ecology which deals with the study of plants in relation with the environment.

Plant Residue: Plant residues which are left in the field following crop harvest are a potentially valuable source of nutrients for succeeding crops.

Plastic Index: The difference between the moisture content at plastic limit and liquid limits is often used as a general index of workability on terms of soil susceptibility to structural damage.

Plastic Limit: It is the state at which soil particle is surrounded by a thin film of water just sufficient to act as lubricant.

Plough sole: A compacted layer at the bottom of the zone of ploughing.

Pluvial rice lands (Rainfed upland): Lands which are well drained without free groundwater within the rooting zone of the rice plant and are generally unbunded.

Podzolization: Podzolization (or Podsolization) is complex soil formation process by which dissolved organic matter and ions of iron and aluminium, released through weathering of various minerals, form organo-mineral complexes (chelates) and are moved from the upper parts of the soil profile and deposit in the deeper parts of soil.

Pollination: The transfer of pollen from anther to a stigma.

Population: A population is an assemblage of individuals of the same species growing or living together simultaneously in one habitat, *i.e.* with respect to time and space/place.

Potassium Absorption Ratio (PAR): A ratio for soil extracts and irrigation waters used to express the relative activity of potassium ions in exchange reactions with soil.

Potassium Deficiency: Symptoms of mild deficiency are dark green leaves, low tillering and stunting. Severe deficiency results in yellowish-orange to yellowish-brown discoloration, starting at the tip of older leaf blades and gradually moving towards base. Necrotic spots may appear on the leaf blades. Grain size and weight may be reduced.

Potential evapo-transpiration: Maximum quantity of water capable of being evaporated in a given climate from a continuous stretch of vegetation covering the whole ground and well supplied with water.

Preparatory Tillage: Various tillage operations carried out on the farm before the crops are sown, planted or transplanted to make a fine seed bed.

Press Mud: Press mud, also known as Filter cake or press cake, is the residual output after the filtration of the muddy juice. It is mixed with spent wash from the distillery and cultivated to produce high quality bio-manure.

Prey Crop (smother crop): The crop which is grown for the purpose of eliminating any undesirable plant through physical suppression or its allelopathic effect.

Primary Growth: Growth which is initiated by apical stem and root meristems.

Primary Index: Term used to denote total sugar (sucrose + reducing sugar) in the sugarcane leaf sheath on dry weight basis. As the ripening starts, PI should be 10 to 11 per cent.

Primary Tillage: Tillage operations which constitute the initial major soil working operations. It is normally designed to reduce soil strength, cover plant materials and rearrange aggregates.

Priming: A term is used to designate the harvesting of ripe tobacco leaves from the stalk as they ripen, beginning at the bottom and progressing upward.

Propping: The operation of tying the leaves together using the bottom dry and green leaves is known as propping. It is done to check lodging of cane. Usually the trash without removing from the cane is twisted to form a sort of rope and cane stalks are tied together.

Protein synthesis: An elaborate process of protein formation in plant cells consisting of activation of amino acids; attachment of amino acids activated to s-RNA and formation of polypeptides on ribosome.

Prussic acid (HCN, Dhurin): A highly toxic gaseous compound, found in certain crops, like sorghum, cassava *etc.* and in some poisonous weeds.

Pseudocereal: A pseudocereal is one of any non-grasses that are used in much the same way as cereals (true cereals are grasses). Their seed can be ground into flour and otherwise used as cereals. Examples of pseudocereals are amaranth

(Love-lies-bleeding, red amaranth, Prince-of-Wales-feather), quinoa, and buckwheat.

Pteridophytes: They are seedless vascular plants, *e.g.* ferns.

Puddle Soil: Dense massive soils artificially compacted when wet and having no regular structure. The conditions commonly result from the tillage of a clay soil when it is wet.

Puddling: Making a soil impermeable by manipulating and compacting it in standing water, which reduces its apparent specific volume that facilitates transplanting.

Pulvimeter: It is used for measurement of Rainfall (RF).

Purity per cent: $= \dfrac{\text{Weight of pure seed}}{\text{Weight of bulk samples (seed+impurities)}} \times 100$

Q_{10}: The number of times the rate of a process (physical, chemical or physiological) increases with a, 10°C rise in temperature. If the rate of process is doubled, Q_{10} is 2. Q_{10} for respiration is 2-3.

Quantum number: The number of molecules of quanta necessary to reduce one molecule of carbon dioxide. It is also called as the quantum requirement.

Quiescence: Dormancy is a state of suspended development of seed/propagule or of any vegetative propagating structure (tuber, bulb, corm or bud). This state is often called as quiescence.

R value: It is the percentage of the crop land actually cropped in a year. Frequency of cropping in a fallow cycle (R) is equal to crop year times 100. R value is synonymous of cropping index.

Rabbing (Tobacco): Process of burning of trashes, weeds or any organic refuse on the soil surface.

***Rabi* (Winter) season:** Season during which winter crops like wheat, barley, mustard, gram, linseed, *etc.* are cultivated and usually it extends from September to April.

Radiation: Transfer of energy from one body to another without the aid of the material medium (solid, liquid or gas).

Radicle: In botany, the *radicle* is the first part of a seedling (a growing plant embryo) to emerge from the seed during the process of germination. The *radicle* is the embryonic root of the plant, and grows downward in the soil.

Rainfall: The total liquid product of precipitation or condensation from the atmosphere as received and measured in a rain gauge in past 24 hours classification of rainfall is as follows:

Very high rainfall – more than 12.5 cm

Heavy rainfall – 6.5 to 12.5 cm

Rather heavy rainfall – 3.5 to 6.5 cm

Moderate rainfall – 0.8 to 3.5 cm

Light rainfall – 0.3 to 0.7 cm

Very light rainfall – upto 0.2 cm

No rainfall – where there is zero rainfall

Rainfed Agriculture: *Rainfed* agriculture is a type of farming that relies on rainfall for water

Rainy day: A recorded rainfall of 2.5 mm or more within a period of 24 hours.

Rancidification: Rancidification is the process of complete or incomplete oxidation or hydrolysis of fats and oils when exposed to air, light, or moisture or by bacterial action, resulting in unpleasant taste and odour.

Ratooning: Ratooning is the agricultural practice of harvesting a monocot crop by cutting most of the above-ground portion but leaving the roots and the growing shoot apices intact so as to allow the plants to recover and produce a fresh crop in the next season.

Real Seed Value: It is the percentage of germinability of Percentage of purity of seed lot of a crop plant.

Recession Farming (*diara* land farming): It is the system in which crops are planted in flooded areas as the rainy season ends and water recedes. This system takes advantage of thoroughly saturated soil profile, and also has the advantage of silt and nutrients left behind by flood water.

Redox Potential (oxygen- reduction potential, Eh): A measure of the tendency of a given oxidation reduction system to donate electrons or to accept electrons. The Eh is determined by measuring the electrical potential difference in volts or millivolts between the given system and standard redox system.

Reduced Tillage: A tillage system in which the primary tillage operation is performed in conjunction with special planting procedure in order to reduce or eliminate secondary tillage operations.

Reference Crop Evapo-Transpiration (ET_0): Rate of evapo-transpiration from an extended surface of 8 to 15 cm tall, green grass cover of uniform height, actively growing, completely shading the ground and not short of water; expressed as mm per day.

Registered Seed: Registered seed means seed produced from plants grown directly from foundation seed, cultivated according to certification regulations and that meets the specifications of the official seed certifying agency of the jurisdiction in which the seed is/was produced.

Regolith: The **regolith** includes all of the weathered material within the soil profile.

Regosol: Any soil of the azonal order without definite genetic horizons and developing from or on deep, unconsolidated, soft mineral deposits such as sands, loess or glacial drift.

Regression Analysis: A technique for the modelling and analysis of numerical data consisting of values of a dependent variable (response variable) and of one or more independent variables (explanatory variable). The dependent variable in the regression equation is modelled as s function of the independent variables, corresponding parameters and a random error term. The parameters are estimated so as to give a "best fit" of the data using the method of least squares.

Relative crowding co-efficient (K): It is the measure of the relative dominance of one component crop to other in a intercropped or mixed cropping system. The co-efficient (K) is determined separately for each component crop, *e.g.* for crop 'a' in association with 'b', the co-efficient is as

$$Kab = \frac{Yab \times Zba}{(Yaa - Yab) \times Zab}$$

where,

Yab- yield of species 'a' in association with 'b'

Yaa- pure stand yield of species 'a'

Zab and Zba- sown proportion of 'a' and 'b' respectively

The product of two co-efficients (Kab×Kba) designated as K, denotes yield advantage if K>1, no advantage if K=1 and disadvantage if k<1.

Relative growth rate (RGR): An index of amount of growing material per unit dry weight of plant per unit time. It is also called efficiency index (y) and can be expressed in $g.g^{-1}.day^{-1}$.

$$= \frac{Log\,W_2 - Log\,W_1}{t_2 - t_1}$$

Where, W_1 and W_2 are total dry weights at time t_1 and t_2, respectively.

Relative net return (intercropping): It is the ratio between the monetary value of base and inter crop plus differential cost of cultivation (intercropping-monoculture of base crop) and monitory value of base crop in monoculture.

Relative yield index: It represents the mean yield for a crop in a region divided by the mean all India yield of the same crop times 100.

Relative yield total: This implies that the mean of plant relative yield (ratio of the per plant yield of the mixture) to that in monoculture will have a value close to unity. In terms of relative yields, the value is a total, which has been called the relative yield total.

Relay cropping: Growing two or more crops simultaneously during the part of life cycle of each. A second crop is planted after the first crop has reached its reproductive stage of growth but before it is ready for harvest. *e.g.* sowing *toria* in standing crop of maize.

Residual Herbicide: Herbicides which after application usually maintain their phytotoxic effect in soil for a considerable period of time, but not for the whole crop growing season. Therefore, they offer good control of weeds in crops for sufficient period at least for the critical period of weed competition.

Residual Sodium Carbonate (RSC): An indicator of the tendency to precipitate Ca as $CaCO_3$ in irrigation water. RSC (me/l) = $(CO_3^{2-} + HCO_3^-) - (Ca^{2+} + Mg^{2+})$. Irrigation water is considered safe if its RCC is < 1.25 and unsafe at RSC > 2.5.

Resilience: The resilience of a crop is the extents of its ability to survive a perturbation or stress with minimum effect on yield. It is, thus, critical component of the crop-yield stability.

In a sole crop, recovery after a shock may involve expansion within between plants of similar genotypes. In an intercrop, compensation is often also between components.

Response Curve: A close functional relationship between the input X (nutrients) and output Y (production) can be expressed in some algebraic equations in the form: Y = f(X)

(i) $Y = a+bX$ (linear response)

(ii) $Y = a+bX+cX^2$ (quadratic response)

(iii) $Y = Y+d\,(1-10^{-kX})$ (Mitscherlich equation)

(iv) $Y = A \times b$ (Cobbs – Douglass or power function)

(v) $Y = ae^{bx}$ (Exponential or logarithmic curve)

Retting: Retting is the process employing the process of microorganisms and moisture on plants to dissolve or rot away much of the cellular tissues and gummy substance surrounding bast fibre bundles and so, facilitating separation of the fibre from the stem.

Reverse Resistance: Reverse resistance is a term used to describe a phenomenon or situation in which weed biotypes resistant to an herbicide fall susceptible to the very herbicide if it is not used for a period of 7-10 years, instead some alternative herbicides used to kill the resistant population. It is believed to be a natural phenomenon that susceptible population becomes resistant on repeated exposure to a herbicide and it reverts back to be susceptible one if the very herbicide is not used for sufficiently long period.

Root Nodules: Refers to small swelling on roots of leguminous plants produced as a result of infection by nitrogen-fixing bacteria (*Rhizobia*).

Rosette: In many plants profuse leaf development with retarded internode growth to give bunchy or rosette appearance.

Rotational Intensity: This is calculated by counting the number of crops grown in a rotation and is multiplied by 100 and then divided by the duration of the rotation.

R-Square: The coefficient of determination is the square of multiple correlation coefficient (correlation between the observed values and the estimated values) and is denoted by R^2. It is obtained as the ratio of the variance explained by the model to the total variation in the data.

Runoff: That portion of precipitation or irrigation on an area which does not infiltrate, but instead is discharged from the area.

Saline: In saline soils, the salinity does not affect the physical properties of soil, but it is harmful because elevated soluble salts in the soil solution reduce the availability of soil water to plants. As soils dry out due to evapotranspiration, the soil water becomes more saline and less available to plants. Saline soils normally have a pH value below 8.5 and have good physical properties. The electrical conductivity (EC) of the soils is generally greater than 4.0 dS/m and the exchangeable sodium percentage (ESP) less than 15. Reclamation of these soils can be accomplished by leaching with high-quality irrigation water. Successful reclamation requires adequate drainage, irrigation water management, and use of the correct amount of water.

Saline Soil: A soil containing excess of neutral soluble salts and pH values of 805, EC 4 dS/m at 25°C and ESP 15 per cent.

Salinization: Salinization is the process by which water-soluble salts accumulate in the soil.

Saltation: Soil movement in wind where particles skip or bounce along the soil surface in response to wind forces. Particles in the size range from 0.1 to 0.5 mm (0.004 to 0.02 in) usually move in this manner.

Sand: Small rock or mineral fragments having diameter ranging from 0.05 – 2 mm.

Sandy Clay: Soil of this textural class contain 35 per cent or more of clay and 45 per cent or more of sand.

Sciophytes: Shade-loving plants are also called sciophytes or sciophytic plants, which are those plants with special ecological adaptation to reduced light intensity or partial sun.

Seasonal Drought: This is found in climates with well-defined rainy and dry season. Most of the arid and semiarid zones fall in this category

Secondary Tillage: The tillage operations following primary tillage to create a good seed bed for proper seeding or planting.

Sedimentary Rocks: Rivers, glaciers and winds carry bits of rock and soil and deposit them in layers. After a few million years, these deposits harden into compact rocks and are called Sedimentary rocks.

Seed Drill: A seed drill is a device that sows the *seeds* for crops by positioning them in the soil and burying them to a specific depth.

Seed Index: Weight of 100 seeds (for bold seeds like maize).

Seed Rate: The required number of plants/unit area is decided by calculating the seed rate. The seed rate depends on spacing or plant population, test weight and germination percentage. It is calculated by the formula.

Seed Rate (kg/ha) = Spacing x Test weight x Germination Percentage x 1,00,000.

Seed Viability: Seed viability is the ability of the embryo to germinate and is affected by a number of different conditions (or) Seed viability is the capability of plant structure (seed, cuttings, *etc.*) to show living properties like germination and growth. or the degree to which the seed is alive (metabolically active).

Seed: Sexually or vegetative propagated planting materials which are used for seeding and planting and as such should be free from any infection and should give a good crop stand by good seeding.

Seed-cum-fertilizer drill: An implement which drills seed and fertilizer simultaneously and uniformly in the same or different rows.

Seedling: The juvenile stage of a plant grown from seed. Usually indicates plants which have up to and including about four true leaves.

Self-pollination: Transfer of pollen from the anther to the stigma of the same flower or another flower on the same plant.

Seminal root: A root arising from the base of hypocotyl as in wheat.

Sensible heat: Sensible heat is heat exchanged by a body or thermodynamic system in which the exchange of heat changes the temperature of the body or system, and some macroscopic variables of the body or system, but leaves unchanged certain other macroscopic variables of the body or system, such as volume or pressure.

Sequential cropping: Sequential cropping refers to growing crops in sequence within a crop year, one crop being sown after the harvest of the other.

Sheet erosion: A form of water erosion in which a very thin layer is removed from the soil surface by detachment and overland flow.

Shelter belt (wind break): A wind barrier of living trees and shrubs established and maintained for protection of crop fields.

Shifting cultivation: The practice of clearing natural vegetation, cultivation for some years, and abandoning them as soon as the soil is exhausted. This includes the practice of shifting homesteads in order to follow the cultivator's search for new fertile land.

Shrinkage limit: It is the state at which the soil passes from moist to dry appearance. Moisture content is just adequate to fill the pores at minimum volume as a result of drying.

Side-Dressing: The application of commercial fertilizer along the side of a row or around a plant.

Siderophile elements: Which tend to form metallic bods.

Siderophores: Siderophores are chelating agents that are secreted by microorganisms and *gramineous* plants in response to iron and zinc deficiency. Under aerated conditions at neutral to alkaline pH, inorganic iron is insoluble. Low molecular weight root exudates dissolve the Fe and make it available to plant.

Sigmoid Growth Curve: A graph sowing the rate of growth of a crop for a period time. Sigmoid means S-shaped. The general pattern is one of initially small increase in size, followed by rapid growth and then by a period during which the size of the crop increases slowly or not at all.

Silage crop: Those crops which are preserved in a succulent condition by partial fermentation in a tight receptacle, *e.g.* sorghum, maize, oat and grapes.

Silage: Animal feed resulting from the storage and fermentation of green or wet crops under anaerobic conditions. Good quality silage has pH 4.5 – 5.5, ammoniacal nitrogen of total N less than 10 per cent, butyric acid < 0.2 per cent, lactic acid 3 – 12 per cent.

Silviculture: Silviculture is the practice of controlling the growth, composition/ structure, and quality of forests to meet values and needs, specifically timber production.

Silvipasture: It is the intentional combination of trees, forage plants and livestock together as an integrated, intensively-managed system.

Simple Correlation: Simple correlation means the strength and direction of a linear association between two random variables. It means the departure of two variables from linear independence. Correlation does not imply causation. (Also refer to correlation coefficient).

Site-Specific Nutrient Management (SSNM) Approach: It involves nutrient application on soil test basis, yield goals, factors influencing crop response to nutrient applications.

Skewness: Skewness is a measure of the asymmetry of the probability of a real-valued random variable.

Smother crop: A dense growing crop which can suppress or stop the growth of a less complicated crop. It is grown for a suppression of weeds, *e.g.* cowpea, mungbean, urdbean *etc.*

Sod culture: A system of soil management where in the plants are grown in permanent grass without tillage.

Sod: Top few cm of soil permeated by and held together with grass roots or grass-legumes roots.

Sodic Soil: Sodic soils have an exchangeable sodium percentage that is high enough to cause a deleterious change in soil flocculation (deterioration in soil aggregation). In extreme cases, sodium ions disperse the mineral colloids, which then form a tight soil structure. This structure slows the infiltration/percolation of water. Irrigation waters containing high amounts of sodium salts versus calcium and/or magnesium salts can create a build-up of exchangeable sodium in the soil. Sodic soils have an EC less than 4.0 dS/m and an ESP greater than 15. Sodic soils normally have a pH greater than 8.5. At very high pH values (>8.5), plant-available P and boron actually increase due to the influence of soluble sodium.

On non-calcareous soils, gypsum or other soluble calcium salts must be applied or elemental S with a liming material, thereby forming gypsum in the soil.

On calcareous soils, treatment may be with acidifying materials that dissolve native calcium, with gypsum. Soluble calcium replaces sodium on the clay surface and improves physical properties that allows sodium and excess salts to be leached or organic materials such as crop residues, manure, and compost may be helpful in providing more soil porosity and better physical condition for leaching.

Soil Adhesion: The sticking of soil to tillage tools or wheels.

Soil Consistence: Resistance of a soil at different soil moisture contents to mechanical stress or manipulations.

Soil Drought: Soil moisture stress due to imbalance between available soil moisture and evapotranspiration of a crop.

Soil Fertility Map: Analysis of the composite samples from each of these grid points is used to make a soil fertility map showing the variation across the field.

Soil Fertility: Soil fertility refers to inherent capacity of a soil to supply nutrients to plants in adequate amounts and in suitable proportions.

Soil Horizon: A soil horizon is a layer parallel to the soil surface whose physical, chemical and biological characteristics differ from the layers above and beneath. Horizons are defined in many cases by obvious physical features, mainly colour and texture.

Soil moisture characteristic curve (moisture retentive curve): A graph showing the relationship between the amount of water remaining in the soil at equilibrium as a function of matric suction.

Soil Pore Space: The pore space (called porosity) found in soil between mineral particles and organic matter is filled with either air or water. The pore space both contains and controls most of the functions of soil. It is not just the total amount of pore space that is importance but also the size and distribution of pores, and the continuity between them that determines function and behaviour of soil. Pore space allows movement of water and air along with the growth of roots.

Soil Productivity: It is the present capacity of the soil to produce crop yield under a defined set of management practices. It is measure in terms of the yield in relation to the input of production factors.

Soil Profile: The *soil* is arranged in layers or horizons during its formation. These layers or horizons are known as the soil profile.

Soil Science: The science dealing with soils as a natural resource on the surface of the earth including soil formation, classification, mapping and physical, chemical, biological and fertility properties of soils present and these properties in relation to their management to crop production.

Soil Sickness: Condition of soil when the productivity is reduced in spite of available nutrients.

Soil Tilth: Tilth refers to the physical condition of the soil in its relation to plant growth. Tilth not only depends on granulation and its stability, but also on moisture content, degree of aeration, rate of water infiltration, drainage, and capillary-water capacity.

Soil-Test Based Crop Recommendations (STCR): Fertilizer recommendation given for a specific field on the basis of available nutrient status of that field as determined by soil analysis. Generally, such recommendation is more balanced and profitable than blanket or general recommendation.

Soil-Water Potential: Soil-water potential, defines water available to plants. It is defined as the amount of work required per unit quantity of water to transport water in soil. The concept of soil-water potential replaces terms such as gravitational, capillary, and hygroscopic water. In the soil, water moves continuously in the direction of decreasing potential energy or from higher water content to lower water content. As a plant takes up water from the soil, the concentration of water in the soil immediately adjacent to its roots is reduced. Water from the surrounding soil then moves into the soil directly around the roots.

Solar Constant: It is the amount of solar energy incident on a unit area at tight angle to the sun's rays at the earth's mean distance per unit time in the absence of the atmosphere. (Solar constant= 1.98 cal/cm^2/min).

Solarization: Solarization uses a sheet of clear plastic film to concentrate the sun's heat and burn out weeds, weed seeds, many plant pathogens such as verticillium wilt, and even pests such as certain nematodes.

Solo Cropping: It is defined as the cultivation of one crop variety alone in pure stand at normal density in a certain time and place.

Sorghum Injury: Poisoning of livestock resulting from feeding of young sorghum plants containing hydrocyanic acid (HCN). HCN concentration is more at seedling stage. Critical concentration of HCN in sorghum is 200 ppm.

Spaced transplanting technique (STP): This technique has been developed for synchronization of tillering and quick seed multiplication of sugarcane.

Specific Crop Intensity Index: It is derivative of Cropping Intensity Index and determines the amount of area-time denoted to each other or group of crops compared to the total time available.

Split Plot Design: A layout in which one set of treatments is assigned to large plots (due to practical requirements it is not possible to change the levels of the treatment) called main plots, in which, a replicate is divided, and another set of treatments to sub-division of the main plots called sub plot, is termed as split-plot design. Each main plot becomes a block for the sub-plot treatments. Split plot design is nothing but a randomized complete block design with restricted randomization. This is also known as factorial experimental with main effects confounded. These designs should only be used when the treatments applied to main plots cannot be allocated to smaller plots of the sizes that of sub-plots.

Split-split plot design: This is an extension of split plot design. Refers to a layout in which one set of treatments is assigned to large plots (due to practical requirements) called main plots, into which, a replicate is divided; second set of treatments is applied to sub-division of the main plots called sub-plots; and third set of treatments are applied to sub-division of sub-plots called sub-sub-plot. Each main plot becomes a block for the sub-ploy treatments and each sub-plot becomes a block for sub-sub-plot treatments. These designs should only be used when the treatments applied to main plots cannot be allocated to smaller plots of the sizes that of sub-plots.

Square root transformation: This is variance stabilizing transformation and is applied when the data consist of small whole numbers and follow a Poisson distribution. For example, data obtained in counting rare events, such as the number of infested plants in a plot, the number of insects caught in traps, number of weeds per plot, *etc.* In such data sets, variance is proportional to mean; square root transformation transforms the data into a normal distribution. If X is the original value, then $y = \sqrt{X}$ represents the square root transformation. If X is confined to small whole numbers then,

$$\sqrt{X + \frac{1}{2}} \text{ or } \sqrt{X + \frac{3}{4}} \text{ should be used instead of } \sqrt{X}$$

Square: An unopened flower bud of cotton with its subtending brackets.

Stable Border: A stable border defines the upwind boundary of an isolated field. It is an area with sufficient protection to prevent saltation from starting, and capable of trapping and holding incoming saltation from eroding areas upwind, thus preventing saltating soil particles from entering areas downwind.

Stale seedbed technique: Stale seedbed technique is a cultural-cum-preventive measure in which irrigation is applied first to the field, then ploughed thoroughly on optimum moisture status and levelled. Crop sowing is withheld and the field is left as such for about a week or so as to allow germination of weeds, which then is controlled by a non-residual herbicide, *e.g.* paraquat, glyphosate, glufosinate-ammonium, or by shallow cultivation with spike-tooth harrow, spring-tooth harrow.

Standard Deviation: A measure of dispersion of variable around the general mean and is calculated as the positive square root of the sum squares of deviations of the observations from the mean value divided by the number of observations.

Standard Error: The positive square root of the variance of the sampling distribution of statistic.

Statistics: It refers to a function of samples values, which does not involve any parameter.

Steenberg's Effect: Under severe deficiency, rapid increase in yield with added nutrient can cause a small decrease in nutrient concentration. It results from dilution of nutrient in the plant by rapid plant growth.

Stefan-Boltzmann Law: One of the radiation laws which states that the amount of energy radiated per unit time from a unit surface area of an ideal black body is proportional to the fourth power of the absolute temperature of the black body; $R = ST^4$.

Sticky Point (soil): The moisture content expressed as a percentage of the oven dried at which needed moist soil just ceases to adhere to external objects. It is a condition of consistency at which the soil barely fails to stick to a foreign object.

Straight fertilizer: Such fertilizers have declarable content of only one major nutrient, *e.g.* Urea, Ammonium sulphate.

Stratification: The arrangement or classification of something into different groups.

Stratopause: The upper boundary of the stratosphere is called the stratopause.

Stratosphere: This layer exists above the tropopause (around 20 km onwards) and extends to altitudes of about 50-55 km.

Strip Cropping: The practice of growing two or more crops in alternating strips along contours to control erosion.

Strip Intercropping: Growing two or more crops simultaneously in different strips wide enough to permit independent cultivation but narrow enough for the crop to interact agronomically.

Strip Plot Design: Two different sets of treatments are tried in large plots with one set of plots superimposed over the other set at right angles. Field is divided into blocks representing replications. Each block of land is further divided into vertical and horizontal strips. First set of treatments are allocated to vertical strips and second set of treatments are allowed to the horizontal strips. Those designs are used when both sets of treatments need larger plots.

Strip Tillage: A tillage system in which only isolated bands of soil are tilled.

Stroke's Law: Velocity of fall of a particle (V) with the same density is directly proportional to the square of the radius (r) and inversely proportional to the viscosity of the medium(e).

Structure: Soil structure is the arrangement of the soil particles into aggregates of various sizes and shapes.

Student-Newman-Keul's test: It is a post-hoc test (or multiple comparison test) and can be used to determine the significant differences between group means in an analysis of variance setting. It is based on the range statistic. In this test, rank all the treatment means in decreasing or increasing order based on the preference of the character under study. The least significant range is obtained by multiplying the standard error of difference of means with

$$\frac{q_{\alpha}(p, edf)}{\sqrt{2}}$$

Where, p is one more than the difference between the ranks of the treatments being compared, *edf* is error degrees of freedom, and the values of $q_{\alpha}(p, edf)$ can be obtained from Studentized Distribution Range Tables.

Sub-Soil Placement: Placement of fertilizer in the sub-soil with the help of heavy power machinery, usually followed in humid and sub-humid regions.

Subsoiling: Any treatment to non-inversively loosen soil below the Ap horizon with a minimum of vertical mixing of the soil. Any treatment to fracture and/or shatter soil with narrow tools below the depth of normal tillage without inversion and with a minimum mixing of the soil. This loosening is usually performed by lifting action or other displacement of soil dry enough so that shattering occurs.

Sulphur Coated Urea (SCU): It has been developed as a slow release nitrogen fertilizer and is basically urea with a coating of elemental sulphur, including a binding agent, a sealant and a mirobicide. The N content of SCU ranges from 10-37 per cent, depending on the thickness of the sulphur coating.

Super Compost: When superphosphate is used during compost making it is called super compost

Superfluous Water: Also called gravitational water. Moisture tension ≤1/3 atm. Such water held at field capacity but not beneficial for higher plants. Harmful for crops if too much.

Surge Irrigation: Surge irrigation is the intermittent application of water used to improve distribution uniformity along a furrow. It works on the principle that dry soil infiltrates water faster than wet soil. When soil is wet is seals because the soil particles at the surface consolidate.

Suspension: Soil movement in wind whereby the finer fractions are transported over long distances floating in the wind stream. Suspension is usually initiated by the impact of saltating particles. Particles moving in this manner are usually less than 0.1 mm (0.004 in) in size. Many suspension-size particles are created by abrasion during erosion.

Sustainable Sugarcane Initiative: It is one of the advanced methods of sugarcane cultivation technique that involves use of less seed material, water, land, labour and optimum utilization of fertilizer to achieve maximum potential yield. SSI is an alternate to conventional method of sugarcane cultivation.

Symbiosis: An intimate physiological association of 2 or more species resulting in mutual benefit, *e.g. Rhizobium* bacteria on the roots of legumes.

Symbiotic Nitrogen Fixation: This process is mediated by bacteria with the ability to convert atmospheric N_2 to plant-available nitrogen while growing in association with a host plant. Symbiotic Rhizobium bacteria fix N_2 in nodules present on the roots of legumes. Through this relationship, the bacteria make N_2 from the atmosphere available to the legume as it is excreted from the nodules into the soil.

Symplast: It constitutes the sum-total of living protoplasm (phloem and other living cells) of a plant. It in other way refers to the interconnected system of living tissues of plant.

Symplastic movement: It is the movement of herbicides through symplast mainly phloem which is normally basipetal (from apex to base). It is a source to sink translocation with sugars or photosynthates.

Synergism: Cooperative action of different chemicals such that the total effect is greater than the sum of independent effects.

Synergistic Effect: A synergistic interaction means the interaction between the two nutrients results in increased yield response that is more than what would be expected of the response to each nutrient individually.

Synoptic Chart: Plotting weather data observations on a large outline map using the international code of signals and abbreviations which in popular term is called as weather map and technically as synoptic chart.

System Approach: Studying a system as an entity made up of all its components and their inter relationships, together with relationship between the system and its environment.

System of Crop Intensification (SCI): This approach seeks not just to get more output from the given amount of inputs, a long standing and universal goal, but aims to achieve higher output with less use of or less expenditure on land, labour, capital and water.

System of Rice Intensification (SRI): The SRI, known as *le system de Riziculture intensive* in French and *la sistema intensive de cultivo Arrocero* (SICA) in Spanish is a climate smart and agroecological methodology for increasing the productivity of irrigated rice by changing the management of plants, soil, water and nutrients. It was developed in 1983 by French Jesuit Father Henri de Laulanie in Madagascar.

Taungya System: A type of agroforestry system in which trees are planted with food crops, which serve as a nursery for the young tree seedlings, the aim is to produce a forestry or tree crop plantation, makes the early years of plantation establishment productive and reduce labour cost of establishment, *e.g.* growing of cassava (tapioca) for initial two to three years in eucalyptus plantation.

Temporary wilting point: Wilting that occurs in hot weather when the rate of transpiration exceeds the rate at which a plant can absorb moisture from the soil. The plant recovers when the temperature falls.

Tensiometer: A tensiometer in soil science is a measuring instrument used to determine the matric water potential.

Ternary fertilizer: Contains three major nutrients *e.g.* Ammonium potassium phosphate.:

Test Weight: It denotes the weight of 1000 grains of field crops. In some crops 100 seed weight also considered.

Texture: Soil texture describes the proportion of the soil particles and the fineness or coarseness of a soil.

The Intergovernmental Panel on Climate Change (IPCC): It is a scientific intergovernmental body under the auspices of the United Nations, set up at the request of member governments. The IPCC is an internationally accepted authority on climate change. Its producing reports have agreements of leading climate scientists and the consensus of participating governments.

Thermo periodism: A phenomenon in which growth or development is promoted by alternating day and night temperatures.

Thermosphere: The thermosphere is the layer in the Earth's atmosphere directly above the mesosphere and below the exosphere.

Thinning: Removing of extra plants from thickly populated crop stand with an idea of maintain optimum plant population.

Tiller: Side-shoot growing from the base of the stem of a cereal or grass plant.

Tipping: When the shoots (primaries) grow in a tea bush after pruning, they are tipped or decapitated at a predetermined height parallel to the ground surface. This operation is called "Tipping". Essentially the initial few rounds of plucking are called tipping.

Tirak: Malformed cotton boll.

Top Dressing: The application of manure is commercial fertilizer to the crop that is already established.

Topography: The arrangement of the natural and artificial physical features of an area

Topping (Tobacco): The removal of the terminal bud with or without some of the small top leaves just before or after the emergence of the flower bud.

Total/Holo-Root Parasite Weeds: Parasite weeds which attach themselves on the host-roots, take away food from the host and do not have any other source of gathering food are total/holo-root parasite weeds, *e.g. Orobanche* sp. on tobacco, tomato, faba bean, chickpea, mustard, *etc.* They are also called obligate root-parasite.

Total/Holo-Stem Parasite Weeds: Parasite weeds which attach themselves on the host-shoots, take away food from the host and do not have any other source of gathering food are total/holo-stem parasite weeds, *e.g. Cuscuta campestris/chinensis/epilinum* (on alfalfa, niger, linseed). They are also called "obligate shoot/stem-parasite."

Trace Elements: The isotope of an element which may be incorporated into a sample to trace the course of the element, alone or in combination, through chemical, biological or physical process.

An old term used for micro-nutrients are minor elements which are taken up in small quantities by a crop plant.

Transpiration: Transpiration consists of the vaporization of liquid water contained in plant tissues and the vapor removal to the atmosphere. Crops predominately lose their water through stomata. These are small openings on the plant leaf through which gases and water vapor pass. The water, together with some nutrients, is taken up by the roots and transported through the plant. The vaporization occurs within the leaf, namely in the intercellular spaces, and the vapor exchange with the atmosphere is controlled by the stomatal aperture. Nearly all water taken up is lost by transpiration and only a tiny fraction is used within the plant.

Transpiration Coefficient: The quantity of water (g) necessary for a plant to produce one kg of dry matter.

Transpiration Pull: Transpiration pull or the suction force is the force which aids in drawing the water upward from roots to leaves.

Transpiration Ratio: The quantity of water transpired to produce a unit amount of dry matter.

Transport: The movement of detached soil material across the land surface or through the air by wind or running water. Transport of soil particles in wind is by three modes: (l) saltation, (2) suspension and (3) surface creep.

Transport Capacity: The maximum amount of soil material that can be carried by wind or running water under given conditions.

Trap Crop: A crop which induces germination of specific parasitic weeds, *e.g. Striga, Orobanche etc.* is called a trap crop. Trap crop is also defined differently as any crop planted in or around a more valuable crop to attract pests away from the latter crop. Later, trap crops are usually heavily treated with pesticides, burnt or ploughed under soil.

Treatment: A single state of some factor being varied in an experiment, such as different fertilizers, dose of fertilizers, method of application of fertilizer, sources of fertilizer, rate of herbicide *etc.*

Tropopause: The top of the troposphere there is a shallow layer separating it from the stratosphere which is known as the "Tropopause".

Troposphere: The lowest region of the atmosphere, extending from the earth's surface to a height of about 6–10 km (the lower boundary of the stratosphere).

True Allelopathy: True allelopathy refers to the release into the environment of chemical compounds that are toxic in the forms they are produced by the plants.

True Density: The mass per unit volume of the particle of a material.

T-Test: Refers to any statistical test in which the test statistic has a Student's t-distribution under the null hypothesis. It is applied when the population is assumed to be normally distributed but the sample size is small enough that the statistic on which inference is based on not normally distributed because it relies on an uncertain estimate of standard deviation rather than on a precisely known value.

Turgidity: Turgidity is the state of being turgid or swollen, especially due to high fluid content.

Turgor pressure: Turgor pressure is the force within the cell that pushes the plasma membrane against the cell wall. It is also called hydrostatic pressure, and defined as the pressure measured by a fluid, measured at a certain point within itself when at equilibrium.

Two-Tailed Test: A test of hypothesis in which the alternative hypothesis is such that it led to two alternatives to the null hypothesis. For instance, testing H_0: Effect of treatment i = Effect of treatment j versus H_1: Effect of treatment i≠ Effect of treatment j leads to a two-tailed test as effect of treatment I may be more or less than the effect of treatment j.

Type I Error: Rejection of null hypothesis when it is true.

Type II Error: Accepting the null hypothesis when it is false.

Ultra-Narrow Row Cotton (UNR): Defined as cotton grown with row spacing of <40 cm. The system may have the potential increase lint yields, decrease risk to losses from water stress and allow the use of conservation system.

Uniformity Trial: An experiment or a set of trials, in which each experimental unit receives exactly the same treatment. The objective, *inter alia,* may be due to some standard characteristics of that treatment or to investigate some aspects of the experimental technique *i.e.* plot shape and size or layout in agricultural trials.

Upland Rice: Rice grown in high level land in rainfed, naturally well-drained soils, without surface-water accumulation, normally without phreatic water supply and normally not bunded.

Urea [$CO(NH_2)_2$]: A synthetic, non-protein organic compound, crystalline or made into granules or prills for fertilizer use and containing 46 per cent nitrogen.

Urea Formaldehyde: Slow release nitrogen fertilizer produced by reaction between urea and formaldehyde.

Urease Inhibitor: Urease inhibitors are used to temporarily reduce the activity of the enzyme and slow the rate at which urea is hydrolysed *e.g.* N-(n-Butyl) triphosphoric triamide (NBTPT), phenylphosphorodiamidate (PPD/PPDA), hydroquinone and ammonium thiosulfate.

Urease: A universally occurring enzyme which activate urea hydrolysis to produce CO_2 and NH_3

Variety: A subdivision are a group of plants within a species which is characterized by growth, plant fruit, seed or other characters by which it can be differentiated from other seeds of the same kind.

Vegetative barrier: These are closely planted grass or shrub rows along the contour for erosion control in agricultural lands. They check the velocity of runoff and retain the sediment by acting as barrier to runoff.

Verification: The evaluation of the accuracy with which the computer codes represent the mathematical model and the programmer's intentions. Verification also involves the careful checking of mathematical manipulations, units and their conversations and programming logic and code to ensure that neither the mathematical model not its translation into one or more computer programs has errors in it.

Vermicompost: A compost produced from the digestion and discharge of refuse and other organic wastes by earthworms.

Vernalization: Low temperature treatment of seeds and seedlings, which has the effect of artificially 'aging' the plant and reducing the time between sowing and flowering.

Vertical mulch: It is developed for heavy soils where infiltration is a problem. Trenches are dug at 5-10 m intervals depending on slope at sizes of 30 x 60 cm across the slope on grade. They are filled with stalk materials which keep the cracks open and allow better water intake.

Vertical Revolution: Maximum production per unit land area per unit of time using intensive cropping systems, high production inputs and improved management practices.

Virgin Soil: A soil that has not been significantly disturbed from its natural environment.

Virtual water: Water deficit countries, facing food shortages, import food grains. These agricultural commodities carry with them embodied water or the water that is needed to grow them. Importing food grains in effect means importing water. Such water is called virtual water.

Volatilization: Volatilization is the loss of nitrogen as ammonia gas (NH_3). There are two major pathways for this loss in agriculture: conversion of ammonium-N (NH_4^+–N) to NH_3 and the conversion of urea ($CO(NH_2)_2$) to NH_3.

Urea is a nitrogen-containing compound that is readily converted to ammonia upon catalysis by the ubiquitous enzyme urease via the following reaction:

$CO(NH_2)_2) + H_2O + \text{urease} —— 2NH_3 + CO_2$

Water Application Efficiency: The percentage of water applied that can be accounted for as moisture increase in the crop root zone of the soil.

Water Infiltration: Water infiltration is the process of water entering the soils from the soil surface. The rate at which water enters the soil, considered either infiltration rate or permeability of the soil, depends on the portion of coarser pores on or near the soil surface. The rate itself is controlled by every factor which affects the number and stability of larger pores. Infiltration rates are directly affected by factors that are somewhat controlled by management including tillage practices, amounts of surface residue, soil organic matter, salinity and sodicity.

Water Productivity (WP): Water productivity is used to define the relationship between crop produced and the amount of water involved in crop production, expressed as crop production per unit volume of water. Crop production may be expressed terms of total dry matter yield of yield or seed (or grain) yield (Kg) or, when dealing with different crops, yield may be transformed into monetary units. It is expressed as kg m^{-3}.

WP = Grain or seed yield/water applied to the field

Water Table: Water tables can be a barrier for root development because of restricted oxygen availability. Most crops grow best where the water table is greater than six feet (15 cm) below the soil surface. Adequate soil drainage must be present for sustained growth of most plants. Additionally, when ground water is saline, upward movement and its subsequent evaporation at the surface of the soil adds to the salinization of soils. Providing artificial drainage for poorly drained soils.

Water logging: Water logging is excess water in the root zone accompanied by anaerobic conditions. The excess water inhibits gaseous exchange with the atmosphere, and biological activity uses up available oxygen in the soil air and water also called anaerobiosis, anoxia or oxygen deficiency.

Watershed: It is a drainage basin or catchment area of a particular stream or river

Watertable: The upper surface of ground-water or that level below which the soil is saturated with water.

Weak Acid Herbicides: Herbicides whose anions or negatively charged forms predominate at neutral pH and have limited adsorption through anion exchange, although are prone to leaching are weak acid herbicides, *e.g.* dinitrophenols, phenylalkanoic ac-ids, phenoxyalkanoic acids, aryloxyphenoxy propionates, chloro-aliphatic acids, benzoic acids, benzoates, sulfonylureas, imida-zolinones, organic arsenicals, triazolopyrimidines, uracils *etc.*

Weak Base Herbicides: Herbicides which form positive charge at acidic soil condition (*i.e.* lower pH) and get adsorbed to soil colloids are weak base herbicides, *e.g.* triazines and triazinones, triazoles, thiadiazoles, pyridinones *etc.*

Weather: The day to day variation in climatic variables/elements such as temperature, humidity, rainfall, sunshine hour *etc.* of a place is called its weather.

Weathering: It is a geological process and essentially destructive in nature which leads to formation of simple compounds from the solid rocks of earth's crust.

Weed Control Efficiency: Weed control efficiency (WCE) denotes the magnitude of weed reduction due to weed control treatment.

Weed Index: Weed index is the measure of the efficiency of a particular treatment when compared with a weed free treatment. It is expressed as percentage of yield potential under weed free.

Weed Persistence Index (WPI): It is calculated as follows:

$$= \frac{\text{Weed population in control plot}}{\text{Weed population in treated plot}} \times \frac{\text{Weed dry weight in treated plot}}{\text{Weed dry weight in control plot}}$$

Weediness: Weediness is the state/condition of a field having abundance of weeds. It implies both presence and population of weeds.

Wet Land: Permitting to soils flooded for several weeks each year, are to crop grown in such soils where the ionic concentrations are expressed in milliequivalents per litre or millimoles per litre.

Wet Year: If the rainfall exceeds twice the normal deviation at a particular place, that year is said to be a wet year.

Wet-Seeded Rice: Sowing of soaked rice seed in standing water (7-15 cm depth) on well prepared seed bed.

Wettable powder (WP): Wettable powder is finely milled powder, which forms a suspension on mixing with water, *e.g.* isoproturon, 2,4-D-Na/K WP formulation. If an active ingredient/herbicide is not soluble in water or organic solvent, it can be ground into fine particles and formulated as a wettable powder containing a high concentration (25-75 per cent) of an active ingredient to which dispersing agents and wetting agents are added to facilitate the dispersion of wettable powder in expanding medium such as water.

Windbreak: A windbreak (shelterbelt) is a planting usually made up of one or more rows of trees or shrubs planted in such a manner as to provide shelter from the wind and to protect soil from erosion. They are commonly planted in hedgerows around the edges of fields on farms.

Winter Killing: Relatively high water transpiration rates in evergreens during a period when absorption of water can proceed only at a relatively slow rate may lead to a type of cold injury called winter killing.

Winterization: The process of removing water from the irrigation system before the onset of freezing temperatures.

Yield Gap: The average gap between potentially achievable crop yield and that which is actually achieved in practice. If reliable crop yield at a research station is 5 t/ha and the average farmer yield is 2 t/ha, the yield gap is 3 t/ha. This represents the untapped realizable potential and level of technology adoption.

Yield Potential: The maximum grain yield of a crop variety is genotype under a given environment where management factors are non-limiting. It is a theoretical value, and upto 80 per cent of this could be achieved with best management.

Yield Stability: The maintenance of yield at desired level over a period of time. Its index is measured by inverse of coefficient of variation, regression between yield and environmental index.

Chapter 5

Essential Elements Functions and Deficiency Symptoms in Crops

Plants need 17 elements for their growth and completion of life cycle (C,H, O, NPK, Ca Mg S, Fe, Mn Zn Cu B Mo Cl Ni)

Classification of Nutrient Elements

1. Basic nutrients: C,H,O
2. Macro-nutrients: N, P, K, Ca, Mg, S
   - ☆ Primary nutrients: N, P, K
   - ☆ Secondary nutrients: Ca, Mg, S
3. Micro-nutrients/trace elements: Fe, Mn, Zn, Cu, B, Mo, Cl, Ni
4. Functional nutrients: Any mineral element that functions in plant metabolism whether or not its action is specific. Nicholas proposed this term in 1961. Na, Co, Vn, and Si are considered as functional nutrients in addition to 17 essential elements.

Nutrient Mobility in Soil

Mobile	*Less Mobile*	*Immobile*
Highly soluble and are not adsorbed on clay complex	Soluble but are adsorbed on clay complex	Highly reactive and get fixed in the soil
NO_3^-, SO_4^{-2}, BO_3^{-3}, Cl^-, Mn^{+2}	NH_4^+, K^+, Ca^+, Mg^{+2}, Cu^{+2}	$H_2PO_4^-$, HPO_3^{-2}, Zn^{+2}

Nutrient Mobility in Plants

Mobile	*Moderately Mobile*	*Less Mobile*	*Immobile*
N,P,K	Zn	S,Fe,Mn,Cu,Mo and Cl	**Ca and B**

Nutrient Functions in Plant and their Deficiency Symptoms

Nitrogen

Functions of Nitrogen

1. Plays an Important role in plant metabolism
2. Being a constituent of chlorophyll, it harnesses solar energy and fixes atmospheric CO_2 as Carbohydrates
3. As amino acids (RNA, DNA), it transfers genetic information and regulates cellular metabolism of amino acids and proteins that form structural units and biological catalysts of phosphorylated compounds involved in energy transformation
4. Major structural constituent of cell and cell wall
5. Increases growth and development of all living tissues and improves the protein content of food grains and quality of fodder, vegetables *etc.*

Deficiency Symptoms of Nitrogen

1. Deficiency symptoms first appear on the older leaves, which then advance towards upper leaves, while the new leaves remain green.
2. A pale yellow chlorosis develops near the tip of the leaf blade and advances towards the base in V shape pattern along the margins until the whole blade turns yellow and then brown, followed by necrosis near the tip.
3. N- deficient plants are generally stunted, develop late and produce less flowers as well as shrivelled grains

Phosphorus

Functions of Phosphorus:

1. Structural component of the membrane system of the cell, chloroplast and mitochondrion
2. It is involved in energy transfer, metabolic processes and basic reactions of photosynthesis, transformation of sugars and starch, nutrient movement in plants.
3. It is necessary for cell division, meristematic growth, root, seed and fruit development.
4. Stimulates flowering, ear emergence and maturation of crops

5. In leguminous crops it hastens and encourages the development of N_2-fixating nodule bacteria

Deficiency Symptoms of Phosphorus

1. Plants fail to make a quick start, develop poor root system, remain stunted
2. Older leaves develop characteristic symptoms of dark to blue-green colouration starting from tips towards the base, which later turns radish brown
3. Under severe deficiency the plants develop spindly growth, reddish-purple in the nodes and internodes due to increase in sugar content and show formation of anthocyanin

Potassium

Functions of Potassium

1. K is involved in opening and closing of stomata, water and nutrient transport and cooling of the plants
2. It activates a number of enzymes, which catalyze the incorporation of amino acids in proteins, synthesis of peptide bonds and phosphate group transfer.
3. It helps in maintaining cytoplasmic pH to 7-8, ideal for many enzymatic reactions
4. It reduces lodging and imparts disease resistance to crops, improves the size of grains of cereals, quality of tobacco, shelf life of fruits and vegetables
5. It plays essential role in the formation of starch and in translocation of sugars.
6. It is of special significance to carbohydrate rich crops *e.g.* sugarcane, potato, sugarbeet *etc.*
7. It increases root growth, and thereby improves tolerance of plants to stress of moisture or drought, heat and lodging

Deficiency Symptoms of Potassium

1. Plant containing less than 1 per cent K is deficient in K
2. Healthy plant tissues generally contain 1-5 per cent K, almost equal to that of N.
3. Its deficiency causes its translocation from older to new leaves, thereby resulting in characteristic visual symptoms on older leaves
4. Symptoms of chlorosis start from leaf margins, followed by scorching and browning of tips and margins in potato, wheat, oat, barley and maize

5. K deficient plants are stunted in growth with pronounced shortening of internodes, and they develop shriveled seed and fruits.

Calcium

Functions of Calcium

1. In plants, Ca forms Ca- pectate, which is an essential constituent of cell wall that maintains the integrity of membranes; and in seeds it is present as Ca-phytate.
2. It is involved in cell division as well as in maintaining chromosome structure
3. Regulates the growth of meristems and functioning of the rot tip and protects the root cells against the ion imbalance, low pH and toxic effect of Al, Fe *etc.*
4. It counteracts the injurious effects of acidic molecules of phosphoric, citric, malic and oxalic acids on plant growth by neutralizing their charges.

Deficiency Symptoms of Calcium

1. Deficiency symptoms appear on younger leaves
2. Plant growth is retarded and younger leaves are short, distorted or torn.
3. Under severe deficiency, the apical meristem dies resulting in the production of tillers from the basal nodes and development of deficiency symptoms on emerging new leaves.
4. In cereals or coarse grain crops the upper internodes are very small and the leaves get crowded to give the appearance of a rosette.
5. Root system is stunted; flowering and maturity is delayed or even barren ears or heads develop.
6. Ca deficiency syndrome in apple are bitter pit
7. Ca deficiency syndrome in tomatoes are blossom end rots

Magnesium

Functions of Magnesium

1. Essential constituent of chlorophyll
2. Regulates the activity of several enzyme systems involved in synthesis of nucleic acids and metabolism of carbohydrates

Deficiency Symptoms of Magnesium

1. Mg deficient plants usually lack vigour and are stunted.
2. As Mg is a mobile nutrient in plants, it deficiency symptoms manifest first on older leaves and advances upwards to younger leaves as interveinal

chlorosis, followed by development of purple lesions within the chlorotic tissues.

3. Cattle grazing on Mg deficient pastures suffer from Grass tetany (nutritional disorder of Mg).

Sulphur

Functions of Sulphur

1. It is an essential constituent of the sulphur containing amino acids.
2. It is a constituent of ferrodoxin containing nitrogenase enzyme, which is involved in biological N-fixation and electron transfer reactions and in the metabolic activities of vitamins, biotin and thiamine.
3. Increases the oil quality in oilseed crops.

Deficiency Symptoms of Sulphur

1. As S is immobile in plants, its deficiency first shows up on the younger leaves as uniform pale green followed by their chlorosis.
2. Oilseeds, pulses and other crops having high requirement of sulphur suffer most from its deficiency.

Micro-nutrients

Iron

Functions of Iron

1. It is necessary for the synthesis and maintenance of chlorophyll in plants and plays an important role in nucleic acid synthesis.
2. Fe activates a number of enzymes, including nitrogenase, aminolevolinic acid synthetase and copro-porphyrinogen oxidase.

Deficiency Symptoms of Iron

1. Fe deficiency is observed in plants growing on calcareous, high pH soils, compact soils with restricted aeration, sandy soils low in organic matter.
2. Fe deficiency shows first on younger leaves.
3. The plants containing <50 mg Fe^{+2}/kg dry matter are categorized deficient in iron.

Manganese (Mn)

Functions of Manganese

1. It is a constituent of water splitting enzyme associated with O2 evolution and electron transfer in photosystem II of photosynthesis.
2. It influences auxin levels and maintain chloroplast membrane structure.

Deficiency Symptoms of Manganese

1. Mn deficiency is observed on plants grown in calcareous soils, leached sandy soils low in organic matter, in upland crops (wheat, berseem) grown in rotation with rice on sandy soils in Punjab.
2. Interveinal chlorosis on the middle and older leaves because of its translocation to younger leaves.
3. Symptoms first appear as very light grey, greyish-yellow or greyish-brown specks of minute size in the interveinal area at the basal part of the leaf which later extend towards the tip.
4. Under severe deficiency, these specks enlarge and coalesce to form spots, streaks, speckles or bands of pinkish-brown or buff colour between the vein that remains green
5. Ultimately the affected areas form necrotic lesions.
6. Plants become stunted and develop weak and restricted root system
7. Syndroms of Mn deficiency are commonly known as
   a. Grey specks of Oat,
   b. Pahala blight of sugarcane,
   c. Speckled yellow of sugarbeet,
   d. Marsh spot of pea.

Zinc

Functions of Zinc

1. Zn deficiency occurs on histosols, sodic, calcareous and sandy soils and on soils wet for prolonged period.
2. Zn-containing enzymes known in higher plants are carbonic anhydrase, superoxide dismutase, RNA polymerase, alcohol dehydrogenase and RNA polymerase
3. Zn play important role in enzyme activation
4. Resistance to disease.
5. Uptake and translocation
6. Zn is also involved in the synthesis of RNA, IAA and metabolism of Gibberellic acid (GA).

Deficiency Symptoms of Zinc

1. In general, the visual symptoms in most of the crop plants develop in the 2nd or 3rd fully mature young leaves.
2. Deficiency symptom is known as white bud in maize or sorghum (white bud is not a bud but rather a band or patch of white tissues)

3. As the deficiency become severe, the tissue develops brown necrosis and purple pigmentation.
4. Internodes and leaves become very short (little leaf) and develop a close cluster to give a rosette appearance, causing stunted plant growth.
5. Zn deficiency causes khaira disease of rice - (symptoms appear on relatively older leaves as small scattered light pale-yellow spots that later enlarges, coalesce and turn deep brown)
6. In citrus Zn deficiency is known as mottle leaf (little leaf) or Frenching

Copper (Cu)

Functions of Copper

1. Constituent of several proteins or enzymes that perform important biochemical functions

Deficiency Symptoms of Copper

1. Cu deficiency is generally observed in crops growing on soils of either inherently low in total Cu or soils high in organic matter (Peat soils).
2. Deficiency symptoms appear on the young mature leaf and even on the emerging one.
3. In cereals, Cu deficiency condition is known as white tip or reclamation disease.
4. Cu deficiency causes senescence, male sterility, delay in flowering and loss of grain yield.
5. In citrus, dieback of new growth results in development of gum pockets between the bark and the wood, commonly known as Exanthema.

Boron

Functions of Boron

1. Transport of sugars
2. Synthesis of cell wall and its structure
3. Lignification's
4. Metabolism of phenols, carbohydrates, IAA, RNA
5. Pollen germination and growth of pollen tube
6. Regulates opening of stomata
7. Imparts drought resistance to crops

Deficiency Symptoms of Boron

1. Deficiency critical in highly calcareous soils, sandy leached soils, limed acid soils or reclaimed yellow or lateritic soils.

2. Boron is immobile in plants
3. Symptoms develop on young leaves as white or transparent lesions in the interveinal mid-areas of the leaf.
4. With the severity of its deficiency, the lesions join together and extend beyond the mid-section while the veins remain green.
5. Under acute deficiency the leaves become thick, curl and become quite brittle, the internodes become short and shoot tip die.
6. Plants or trees form bushy appearance.
7. Plants develop poor root growth, defective inflorescence and even flowers do not form.
8. The deficiency syndromes are known as:-
    a) brown heart or heart rot in Sugarbeet and Turnip
    b) browning of Curds and hollow stem in Cauliflower
    c) External and internal cork in Apple

Molybdenum (Mo)

Functions of Molybdenum

1. Essential constituent of nitrogenase enzymes that catalysis biological nitrogen fixation by root-nodules bacteria of legume plants
2. Essential constituent of nitrate reductase enzyme that reduces nitrate to nitrite
3. Essential constituent of xanthine oxidase/dehydrogenase which catalysis the catabolic pathway of purines to uric acid.

Deficiency Symptoms of Molybdenum

1. Mo deficiency is observed in plants growing in acid soils.
2. Deficiency symptoms appears on old or middle leaves as interveinal chlorotic mottling, followed by marginal necrosis and in folding of leaves.
3. Under severe deficiency, mottled area becomes necrotic, causing leaf tissues wither from margins towards the lamina and midrib leaving only the midrib, and a few small pieces of lead blade, which give the appearance of whip or tail.
4. Whip tail of cauliflower and broccoli is caused by Mo deficiency.

Chlorine (Cl)

Functions of Chlorine

1. Regulates water potential of plant through osmotic adjustment *viz.* stomata opening, cell elongation, turgor pressure and osmotic pressure.

2. It acts as a co-factor in water splitting Mn-enzymes
3. It provides resistance to plants against several diseases.

Deficiency Symptoms of Chlorine

1. Irregular yellow to necrotic spotting of leaves, premature wilting, chlorosis of newly emerging leaves, and reduced shoot and root growth.

Nickel (Ni)

Functions of Nickel

1. It is essential for hydrogenase, methyl reductase and urease activities that regulate N metabolism.
2. It is needed for grain filling and seed vitality.

Deficiency Symptoms of Nickel

1. Its deficiency causes accumulation of nitrates and decrease in amino acids content in barley.

Chapter 6

Weed Management

Principles

The selectivity exhibited by certain chemicals to cultivated crops in controlling its associated weeds without affecting the crops forms basis for the chemical weed control. Such selectivity may be due to differences in the morphology, differential absorption, differential translocation, differential deactivation *etc.*

Timeline of Developments in Weed Management

Between 15000 and 10000 BC	Primitive agriculture was begun by Neolithic man
6500 BC	Development of irrigation along the Euphrates and Tigris rivers in Mesopotamia and the Nile river in Egypt
3000 BC	Bone or wood-made hand sickle was invented
1000 BC	Introduction of Tshaped wooden implement pulled by cattle or slaves
164 BC	Romans brought down Carthage using sulfuric acid, salts of copper and arsenic, and carbon bisulfide at high rates to destroy their agriculture
100	Importance of weeding was realized by the Romans
1617	The English developed wood plough with an iron pointed shear
1722	Horse drawn 'hoe' developed by Jethro Tull in Britain
1777	Use of the plough as the chief means of weed control in Sweden

1797	Thomas Jefferson patented a mouldboard plough with iron shear
1840	Lime was recommended for control of horsetail (*Equisetum* sp.) in Germany
1854	Salt (sodium chloride) as a herbicide was recommended in Germany
1848	Wheel cultivator with steel shovel
1850	Self propelled steam tractor invented
1850	Hand sprayer developed and used for insect and disease control
1855	Kirchhof recommended sulfuric acid and iron sulfate against weeds in Germany
1887	Tractor sprayer developed
1896	Copper sulfate was first used for selective weed control of charlock (*Brassica kaber*) in wheat by the French
1897-1900	Copper sulfate, iron sulfate, sulfuric acid, and nitric acid were tested for control of annual forbs in cereals by other workers in France, Germany, Canada, and the United States
1905	Copper sulfate was introduced as algicide in the United States
1900	Gasoline-powered sprayer was developed
1900-37	Gasoline powered crusher boats were used extensively by the US Army Corps of Engineers for control of water hyacinth in Louisiana
1902-37	Sodium arsenite was used extensively by the Army Corps of Engineers for control of water hyacinth in Louisiana
1906	Carbon bisulfide was found very effective as a soil fumigant for weed control
1908	Bolley used various salts for non-selective control in wheat fields
1912-14	Rod weeder, first field implement designed exclusively for weed control was developed in the Pacific Northwest
1914	Orchard heating oil was used for control of garlic (*Allium tricoccum*) and wild onions (*Allium canadense*)
1913-45	Sodium arsenite was used in Hawaii for controlling annual weeds
1920	The tractor-mounted cultivator, the duck foot and blade cultivators were developed and utilized for weed management
1925	The air blast mist blower was developed
1926	Jet-type pumps were developed
1926	Sodium arsenite was found highly effective on submersed aquatic weeds in Wisconsin lakes
1930	Tractor-mounted sprayer was developed
1932	The first organic herbicide, dinitro-ortho cresol (DNOC) was introduced

1933	Salt of dinitrophenol was used in France to control annual forbs in cereals
1937	Use of sodium arsenate in Punjab, India for the control of *Carthamus oxycantha*
1937-50	Common salt was used extensively for field bindweed (*Convolvulus arvensis*) control on highways and railroad in Kansas
1940	Introduction of granule spreader for applying sodium chlorate as herbicide
1941	Synthesis of 2,4-D
1944	Selective nature of 2,4-D was discovered on dandelion
1945	U.S. Patent No. 2,390,941 was issued for 2,4-D to plant physiologist Dr. Franklin and D. Jones of the American Chemical Paint Company
1946	2,4-D was registered for use in Canada on crops and turf grass
1948	The effectiveness of emulsified xylene-type aromatic solvents for control of submerged aquatic weeds in flowing water was discovered
1951	Phenyl substituted urea's (monuron, diuron) were discovered
1955	Triazines (atrazine, simazine) were developed
1960	Dinitroanilines (trifluralin, oryzalin) were introduced
1965	Agent Orange was first introduced in South Vietnam when it was used to defoliate portions of the banks of the Saigon river from the capital city to the South China Sea (Cecil 1986)
1970	First documented case of herbicide resistant biotypes of weeds appeared
1970	Glyphosate was developed
1980	ALS inhibitors (imidazolines, sulfonyl urea's) were developed

Sources: Stainmaus, 2008; Holm and Johnson, 2009; Das, 2014.

Timeline of Herbicides since 1936

Herbicide	***Chemical Nomenclature***	***Year***
Pentachlorophenol	2,3,4,5,6-Pentachlorophenol	1936
2,4,5-T	(2,4,5-Trichlorophenoxy) acetic acid	1944
MCPA	(4-chloro-2-Methylphenoxy) acetic acid	1945
Propham	Isopropyl Carbanilate	1945
2,4-D	2,4-Dichlorophenoxyacetic acid	1946
TCA	Trichloroacetic acid	1947
Fenuron	1,1-Dimethyl-3-phenylurea	1951

Herbicide	*Chemical Nomenclature*	*Year*
Neburon	1-Butyl-3-(3,4-dichlorophenyl)-1-Methylurea	1951
Monuron	3-(p-Chlorophenyl)-1,1-Dimethylurea	1952
2,3,6-TBA	2,3,6-Trichlorobenzoic Acid	1952
Endothall	7-Oxabicyclo (2.2.1) heptane-2,3-dicarboxylic acid	1953
Metham	Sodium methyldithiocarbamate	1953
Mecoprop	(RS)-2-(4-Chloro-2-Methylphenoxy) propanoic acid	1953
Dalapon	2,2-Dichloropropionic acid	1954
Diuron	3-(3,4-Dichlorophenyl)-1, 1-dimethylurea	1954
Silvex	2-(2,4,5-Trichlorophenoxy) Propionic acid	1954
MCPB	4-(4-chloro-2-Methylphenoxy) butanoic acid	1955
Amitrole	3-Amino-s-triazole	1955
Erbon	2-(2,4,5-Trichlorophenoxy) ethyl 2,2-dichloropropionate	1955
Naptalam	N-1-Naphthylphthalamic acid	1955
Simazine	2-chloro-4,6-bis (Ethylamino)-s-triazine	1956
CDAA	N,N-Diallyl-2-chloroacetamide	1956
Chlorpropham	Isopropyl-m-chlorocarbanilate	1956
DSMA	Disodium methanearsonate	1956
Monuron TCA	3-(p-Chlorophenyl)-1,1-dimethylureamono (trichloroacetate)	1956
Sesone	2-(2,4-Dichlorophenoxy) ethyl sodium sulfate	1956
Propazine	6-Chloro-N_2,N_4-diisopropyl-1,3,5-triazine-2,4-diamine	1957
CDEC	2-Chloroallyl diethyldithiocarbamate	1957
Dazomet	Tetrahydro-3,5-dimethyl-5H-1,3,5-thiadiazine-2-thione	1957
EPTC	S-Ethyl dipropylthiocarbamate	1957
HCA	1,1,1,3,3,3-Hexachloro-2-propanone	1957
PBA	Chlorinated benzoic acid	1957
Paraquat	1,1'- Dimethyl-4,4'-bipyridinium dichloride	1958
2,4-DB	4-(2,4-Dichlorophenoxy) butyric acid	1958
Trietazine	6-Chloro-N_2, N_2, N_4-triethyl-1,3,5-triazine-2,4-diamine	1958
Pebulate	S-Propyl butyl(ethyl)thiocarbamate	1959

Herbicide	***Chemical Nomenclature***	***Year***
Atrazine	2-Chloro-4-(ethylamino)-6-(isopropylamino)-s-triazine	1959
Fenuron TCA	1,1-Dimethyl-3-phenylurea monotrichloroacetate	1959
DCPA	Dimethyl tetrachloroterephthalate	1959
Vernolate	S-Propyl dipropylthiocarbamate	1959
Chlorthal	(RS)-2-Chloro-5-(1-hydroxy-3-oxo-2,3-dihydro-1H-isoindol1-yl)benzene-1-sulfonamide	1960
Propanil	N-(3,4-Dichlorophenyl) propanamide	1960
Diphenamid	N, N-dimethyldiphenylacetamide	1960
Acrolein	Acrolein or 2-propenal	1960
Amiben	3-Amino-2,5-dichlorobenzoic acid	1960
Diallate	S-(2,3-dichloroallyl) Diisopropylthiocarbamate	1960
Fenac	(2,3,6-Trichlorophenyl) acetic acid	1960
2,4-DEP	Tris[2-(2,4-dichlorophenoxy) ethyl] phosphate	1960
Triallate	S-2,3,3-Tricloroallyl di-isopropylthiocarbamate	1960
Prometon	N_2, N_4-diisopropyl-6-methoxy-1,3,5-triazine-2	1960
Pentanochlor	3'-Chloro-2-methylvaler-p-toluidide	1960
Dichlorprop (2,4-DP)	(R)-2-(2,4-Dichlorophenoxy) propionic acid	1961
Amitrole-T	3-Amino-s-triazole+ammonium thiocyanate	1961
Barban	4-Chloro-2-butynyl-m-chlorocarbanilate	1961
Cacodylic	Acid hydroxydimethylarsine oxide	1961
Propazine	2-Chloro-4,6-bis(isopropylamino)-s-triazine	1961
Trifluralin	2,6-Dinitro-N, N-dipropyl-4-(trifluoromethyl) aniline	1961
Bensulide	S-2-Benzenesulfonamidoethyl O, O -diisopropyl phosphorodithioate	1962
Prometryn	6-Methylsulfanyl-2-N,4-N-di(propan-2-yl)-1,3,5-triazine-2,4-diamine	1962
Linuron	3-(3,4-Dichlorophenyl)-1-methoxy-1-methylurea	1962
Chloridazone	5-Amino-4-chloro-2-phenylpyridazin-3(2H)-one	1962
Desmetryn	N_2-Isopropyl-N4-methyl-6-methylthio-1,3,5-triazine-2,4-diamine	1962
Terbacil	3-Tert-butyl-5-chloro-6-methyluracil	1962
Dicamba	3,6-DIchloro-o-anisic acid	1962
Dichlobenil	2,6-Dichlorobenzonitrile	1962

Herbicide	***Chemical Nomenclature***	***Year***
Diquat	6,7-Dihydrodipyrido[1,2-a:2',1'-c] pyrazinediium salts	1962
Isocil	5-Bromo-3-isopropyl-6-methyluracil	1962
Molinate	S-Ethylhexahydro-1H-azepine-l-carbothioate polychlorodicyclopentadiene	1962
Triallate	S-(2,3,3-Trichloroallyl) diisopropylthiocarbamate	1962
Monolinuron	3-(4-Chlorophenyl)-1-methoxy-1-methylurea	1962
Carbetamide	(R)-1-(Ethylcarbamoyl)ethyl carbanilate	1963
Ioxynil	4-Hydroxy-3, 5-diiodophenyl cyanide	1963
Picloram	4-Amino-3,5,6-trichloro-2-pyridinecarboxylic acid	1963
Benfluralin	N-Butyl-N-ethyl-2,6-dinitro-4-(trifluoromethyl) aniline	1963
Bromacil	5-Bromo-3-sec -butyl-6-methyluracil	1963
Butylate	S-Ethyl diisobutylthiocarbamate	1963
Diphenamid	N,N-Dimethyl-2,2-diphenylacetamide	1963
MSMA	Monosodium methanearsonate	1963
Picloram	4-Amino-3,5,6-trichloropicolinic acid	1963
Trifluralin	α, α,α-Trifluoro-2,6-dinitro-N,N-dipropyl-p-toluidine	1963
Benazolin	4-Chloro-2,3-dihydro-2-oxo-1,3-benzothiazol-3-ylacetic acid	1964
Siduron	1-(2-methylcyclohexyl)-3-phenylurea	1964
Lenacil	3-Cyclohexyl-1,5,6,7-tetrahydrocyclopentapyrim idine2,4(3H)-dione	1964
Propachlor	2-Chloro-N-isopropyl-N-phenylacetamide	1964
Fluometuron	1,1-Dimethyl-3-(3-(trifluoromethyl) phenyl) urea	1964
Ametryne	2-(Ethylamino)-4-(isopropylamino)-6(methylthio)-s-triazine	1964
Cycloate	S-Ethyl N-ethylthiocyclohexanecarbamate	1964
Molinate	S-ethyl azepane-1-carbothioate	1964
Prometryne	2,4-bis(isopropylamino)-6-(methylthio)-s-triazine	1964
Vernolate	S-propyl dipropyl(thiocarbamate)	1964
Asulam	N-(4-aminophenyl) sulfonylcarbamic acid methyl ester	1965
Terbutryn	N_2-tert-butyl-N_4-ethyl-6-methylthio-1,3,5-triazine-2,4-diamine	1965

Herbicide	***Chemical Nomenclature***	***Year***
Dinoterb	2-(2-Methyl-2-propanyl)-4,6-dinitrophenol	1965
Monalide	4'-Chloro-a, a-dimethylvaleranilide	1965
Bromoxynil	3,5-Dibromo-4-hydroxybenzonitriole	1965
Pyrazon	5-Amino-4-chloro-2-phenyl-3(2H)-pyridazinone	1965
Methoprotryne	N_2-Isopropyl-N_4-(3-methoxypropyl)-6-methylthio-1,3,5-triazine-2,4-diamine	1965
Chlorbromuron	3-(4-bromo-3-chlorophenyl)-1-methoxy-1-methylurea	1966
Terbumeton	N_2-Tert-butyl-N_4-ethyl-6-methoxy-1,3,5-triazine-2,4-diamine	1966
Benefin	N-Butyl-N-ethyl-a, a, a-trifluoro-2,6-dinitro-p-toluidine	1966
Chloroxuron	3-[p-(p-Chlorophenoxy) phenyl]-1,1-dimethylurea	1966
Nitrofen	2,4-Dichlorophenyl p -nitrophenyl ether	1966
Terbacil	3-Tert-butyl-5-chloro-6-methyluracil	1966
Terbuthylazine	N_2-tert-butyl-6-chloro-N_4-ethyl-1,3,5-triazine-2,4-diamine	1966
Benzthiazuron	1-(1,3-benzothiazol-2-Pheyl)-3-methylurea	1967
Phenmedipham	Methyl 3-(3-methylcarbaniloyloxy) carbanilate	1967
Nitralin	4-(Methylsulfonyl)-2,6-dinitro-N, N-dipropylaniline	1967
Bentazone	3-Isopropyl-1H-2,1,3-benzothiadiazin-4(3H)-one 2,2-dioxide	1968
Dipropetryn	6-Ethylthio-N_2, N_4-diisopropyl-1,3,5-triazine-2,4-diamine	1968
Metoxuron	3-(3-Chloro-4-methoxyphenyl)-1,1-dimethylurea	1968
Methabenzthia-zuron	1-(1,3-Benzothiazol-2-yl)-1,3-dimethylurea	1968
Metribuzin	4-Amino-6-tert-butyl-4,5-dihydro-3-methylthio-1,2,4-triazine 5-one	1968
Norflurazon	4-Chloro-5-methylamino-2-(α, α, α-trifluoro-m-tolyl)pyridazin-3(2H)-one	1968
Karbutilate	(3,3-Dimethylureido) phenyl-tert-butylcarbamate	1968
Chlortoluron	3-(3-Chloro-4-methylphenyl)-1,1-dimethylurea	1969
Napropamide	(RS)-N, N-diethyl-2-(1-naphthyloxy) propionamide	1969
Flupropanate	2,2,3,3-Tetrafluoropropionic acid	1969

Herbicide	***Chemical Nomenclature***	***Year***
Bromofenoxim	3,5-Dibromo-4-hydroxybenzaldehyde-2,4-dinitrophenyloxime	1969
Desmedipham	ethyl-3-phenylcarbamoyloxycarbanilate	1969
Ethofumesate	(2-Ethoxy-3,3-dimethyl-2H-1-benzofuran-5-yl) methane-sulfonate	1969
Thiobencarb	S-4-Chlorobenzyl diethyl(thiocarbamate)	1969
Oxadiazon	5-Tert-butyl-3-(2,4-dichloro-5-isopropoxyphenyl)-1,3,4-oxadiazol-2(3H)-one	1969
Propyzamide	3,5-Dichloro-N-(1,1-dimethylprop-2-ynyl) benzamide	1969
Butachlor	N-(Butoxymethyl)-2-chloro-N-(2,6-diethylphenyl) acetamide	1970
Butralin	(RS)-N-sec-butyl-4-tert-butyl-2,6-dinitroaniline	1971
Glyphosate	N-(Phosphonomethyl) glycine	1971
Perfluidone	1,1,1-Trifluoro-N-(4-phenylsulfonyl-o-tolyl) methane sulfonamide	1971
Fluchloralin	N-(2-Chloroethyl)-2,6-dinitro-N-propyl-4-(trifluoromethyl) aniline	1972
Piperophos	S-2-Methylpiperidinocarbonylmethyl O, O-dipropyl phosphorodithioate	1972
Quinoclamine	2-Amino-3-chloro-1,4-naphthoquinone	1972
Bifenox	Methyl-5-(2,4-dichlorophenoxy)-2-nitrobenzoate	1973
Chloramben	3-Amino-2,5-dichlorobenzoic acid	1973
Ethidimuron	1-(5-Ethylsulfonyl-1,3,4-thiadiazol-2-yl)-1,3-dimethylurea	1973
Oryzalin	3,5-Dinitro-N_4, N_4-dipropylsulfanilamide	1973
Profluralin	N-(Cyclopropylmethyl)-2,6-dinitro-N-propyl-4-(trifluoromethyl)aniline	1973
Thiazafluron	1,3-Dimethyl-1-(5-trifluoromethyl-1,3,4-thiadiazol-2-yl) urea	1973
Dimefuron	3-[4-(5-Tert-butyl-2,3-dihydro-2-oxo-1,3,4-oxadiazol-3-yl)-3-chlorophenyl]-1,1-dimethylurea	1974
Ethalfluralin	N-Ethyl-trifluoro-N-(2-methylallyl)-2,6-dinitro-ptoluidine	1974
Tebuthiuron	1-(5-Tert-butyl-1,3,4-thiadiazol-2-yl)-1,3-dimethylurea	1974
Hexazinone	3-Cyclohexyl-6-dimethylamino-1-methyl-1,3,5-triazine-2,4-dione	1974

Herbicide	Chemical Nomenclature	Year
Flamprop	Methyl N-benzoyl-N-(3-chloro-4-fluorophenyl)-D-alaninate	1974
Thiocarbazil	S-benzyl di-sec -butyl(thiocarbamate)	1974
Metolachlor	(RS)-2-Chloro-N-(2-ethyl-6-methyl-phenyl)-N-(1-methoxypropan-2-yl) acetamide	1974
Pyridate	O-6-Chloro-3-phenylpyridazin-4-yl-S-octyl thiocarbonate	1974
Pendimethalin	N-(1-ethylpropyl)-2,6-dinitro-3,4-xylidine	1974
Fosamine	Ethyl hydrogen carbamoylphosphonate	1974
Butamiphos	O-ethyl-O-6-nitro-m-tolyl sec-butylphosphoramidothioate	1975
Diclofop	4-(2,4-Dichlorophenoxy) phenol	1975
Orbencarb	S-2-Chlorobenzyl diethyl(thiocarbamate)	1975
Metamitron	4-Amino-4,5-dihydro-3-methyl-6-phenyl-1,2,4-triazin-5-one	1975
Oxyfluorfen	2-Chloro-trifluoro tolyl-3-ethoxy-4-nitrophenyl ether	1975
Daimuron	1-(1-Methyl-1-phenylethyl)-3-p-tolylurea	1975
Triclopyr	[(3,5,6-trichloro -2-pyridinyl) oxy] acetic acid	1975
Flamprop-M	N-benzoyl-N-(3-chloro-4-fluorophenyl)-D-alanine	1976
Dimethachlor	2-Chloro-N-(2-methoxyethyl) acet-2',6'-xylidide	1977
Fluridone	1-Methyl-3-phenyl-5-[3-(trifluoromethyl) phenyl] pyridin-4-one	1977
Tebutam	N-Benzyl-N-isopropyl-2,2-dimethylpropionamide	1977
Acifluorfen	5-[2-Chloro-4-(trifluoromethyl) phenoxy] -2-nitrobenzoic acid	1980
Benfuresate	2,3-Dihydro-3,3-dimethylbenzofuran-5-yl ethanesulfonate	1980
Chlorsulfuron	1-(2-Chlorophenylsulfonyl)-3-(4-methoxy-6-methyl-1,3,5-triazin-2-yl) urea	1980
Isouron	3-(5-tert-butyl-1,2-oxazol-3-yl)-1,1-dimethylurea	1980
Pyrazolynate	4-(2,4-dichlorobenzoyl)-1,3-dimethylpyrazol-5-yl toluene-4-sulfonate	1980
Anilofos	S-4-chloro-N-isopropylcarbaniloylmethyl O, O-dimethyl phosphorodithioate	1981
Fluazifop-P-butyl	Butyl (R)-2-{4-[5-(trifluoromethyl)-2-pyridyloxy] phenoxy} propionate	1981

Herbicide	*Chemical Nomenclature*	*Year*
Glufosinate	(2RS)-2-amino-4-[hydroxy(methyl)phosphinoyl] butyricacid	1981
Dinitramine	N_1, N_1-diethyl-2,6-dinitro-4-trifluoromethyl-m-phenylenediamine	1982
Flurochloridone	Mixture of the enantiomeric pairs (3RS,4RS;3RS,4SR)-3-chloro-4-chloromethyl-1-(α, α,α -trifluoro-m-tolyl)-2-pyrrolidone (isomers in the ratio 3:1)	1982
Imazamethabenz	Reaction mixture of 6-[(RS)-4-isopropyl-4-methyl-5-oxo-2-imidazolin-2-pheyl-toluic acid and 2-[(RS)-4-isopropyl-4-methyl-5-oxo-2-imidazolin-2-pheyl]-p-toluic acid in the ratio 3:2	1982
Metazachlor	2-Chloro-N-(pyrazol-1- methyl) acet-2',6'-xylidide	1982
Sulfometuronmethyl Methyl	2-(4,6-Dimethylpyrimidin-2-ylcarbamoylsulfamoyl) benzoate	1982
Imazapyr	2-[(RS)-4-Isopropyl-4-methyl-5-oxo-2-imidazolin-2-yl] nicotinic acid	1983
Imazaquin	2-[(RS)-4-Isopropyl-4-methyl-5-oxo-2-imidazolin-2-yl] quinoline-3-carboxylic acid	1983
Quizalofop	(RS)-2-[4-(6-Chloroquinoxalin-2-yloxy) phenoxy] propionic acid	1983
Aclonifen	2-chloro-6-nitro-3-(phenoxy)aniline	1983
Bensulfuron-methyl	Methyl -(4,6-dimethoxypyrimidin-2-ylcarbamoylsulfamoyl)-o-toluate	1983
Metsulfuron-methyl	2-(4-Methoxy-6-methyl-1,3,5-triazin-2-yl) amino]-oxomethyl] sulfamoyl} benzoic acid methyl ester	1983
Fomesafen	5-(2-Chloro- α, α-trifluoro-p-tolyloxy)-N-mesyl-2-nitrobenzamide	1983
Sethoxydim	(5RS)-2-[(EZ)-1-(Ethoxyimino) butyl]-5-[(2RS)-2-(ethylthio) propyl]-3-hydroxycyclohex.	1983
Clopyralid	3,6-Dichloro-2-pyridinecarboxylic acid	1984
Pyrazoxyfen	2-[4-(2,4-Dichlorobenzoyl)-1,3-dimethylpyrazol-5-yloxy] acetophenone	1984
Dimepiperate	S-1-Methyl-1-phenylethyl piperidine-1-carbothioate	1984
Isoxaben	N-[3-(1-Ethyl-1-methylpropyl)-1,2-oxazol-5-yl]-2,6-dimethoxybenzamide	1984

Herbicide	Chemical Nomenclature	Year
Imazethapyr	5-Ethyl-2-[(RS)-4-Isopropyl-4-methyl-5-oxo-2-imidazolin-2-yl] nicotinic acid	1984
Acetochlor	2-Chloro-N-(ethoxymethyl)-N-(2-ethyl-6-methylphenyl) acetamide	1985
Cycloxydim	(RS)-2-[(EZ)-1-(ethoxyimino) butyl]-3-hydroxy-5-[(3RS)-thian-3-yl] cyclohex-2-en-1-one	1985
Thifensulfuron	1,3-Dimethyl-1-(5-trifluoromethyl-1,3,4-thiadiazol-2-yl) urea	1985
Fluroxypyr	[(4-Amino-3,5-dichloro-6-fluoro-2-pyridinyl) oxy] acetic acid	1985
Quinclorac	3,7-Dichloro-8-quinolinecarboxylic acid	1985
Triasulfuron	1-[2-(2-Chloroethoxy) phenylsulfonyl]-3-(4-methoxy-6-methyl-1,3,5-triazin-2-yl) urea	1985
Quinmerac	7-Chloro-3-methylquinoline-8-carboxylic acid	1985
Tribenuron-methyl	Methyl-2-[4-methoxy-6-methyl-1,3,5-triazin-2-yl(methyl) carbamoyl sulfamoyl] benzoate	1985
Chlorimuron-ethyl	Ethyl 2-[(4-chloro-6-methoxypyrimidin-2-yl) amino] carbonyl] amino] sulfonyl] benzoate	1986
Clomazone	2-[(2-chlorophenyl) methyl]-4,4-dimethyl-3-isoxazolidinone	1986
Fluxofenim	4'-Chloro-2,2,2-trifluoroacetophenone (EZ)-O-1,3-dioxolan-2-ylmethyloxime	1986
Clethodim	(5RS)-2-{(E)-1-[(2E)-3-Chloroallyloxyimino] propyl}-5-[(2RS)-2-(ethylthio) propyl]-3-hydroxycyclohex-2-en-1-one	1987
Cinosulfuron	1-(4,6-Dimethoxy-1,3,5-triazin-2-yl)-3-[2-(2-ethoxyethoxy) phenyl sulfonyl] urea	1987
Flurtamone	(RS)-5-Methylamino-2-phenyl-4-(α, α-trifluoro-m-tolyl) furan-3(2H)-one	1987
Lactofen	Ethyl-O- [5-(2-chloro- α, α-trifluoro-p-tolyloxy)-2-nitrobenzoyl]-DL-lactate	1987
Mefenacet	2-(1,3-benzothiazol-2-yloxy)-N-methylacetanilide	1987
Primisulfuron-methyl	Methyl2-[4,6-bis(difluoromethoxy)pyrimidin-2-ylcarbamoylsulfamoyl] benzoate	1987
Prodiamine	5-Dipropylamino- α, α-trifluoro-4,6-dinitro-o-toluidine	1987
Propaquizafop	2-Isopropylideneaminooxyethyl-(R)-2-[4-(6-chloroquinoxalin-2-yloxy) phenoxy] propionate	1987

Herbicide	***Chemical Nomenclature***	***Year***
Tralkoxydim	(RS)-2-[(EZ)-1-(ethoxyimino) propyl]-3-hydroxy-5-mesitylcyclohex	1987
Benzofenap	2-[4-(2,4-Eichloro-m-toluoyl)-1,3-dimethylpyrazol-5-yloxy]-4'-methylacetophenone	1987
Fluoroglycofen	O-[5-(2-Chloro- trifluoro-p-tolyloxy)-2-nitrobenzoyl] glycolic acid	1987
Isoproturon	3-(4-Isopropylphenyl)-1,1-dimethyl urea	1987
Esprocarb	S-Benzyl (RS)-1,2-dimethylpropyl(ethyl) thiocarbamate	1988
Prosulfocarb	S-Benzyl dipropyl thiocarbamate	1988
Clomeprop	(RS)-2-(2,4-dichloro-m-tolyloxy) propionanilide	1989
Cinmethylene	(1RS,2SR,4SR)-1,4-epoxy-p-menth- 2-yl 2-methylbenzyl ether	1989
Flazasulfuron	1-(4,6-Dimethoxypyrimidin-2-yl)-3-(3-trifluoromethyl-2-pyridylsulfonyl) urea	1989
Isoxapyrifop	(RS)-2-[2-[4-(3,5-dichloro-2-pyridyloxy) phenoxy] propionyl]-1,2-oxazolidine	1989
Pyrazosulfuronethyl	Ethyl-5-[(4,6-dimethoxypyrimidin-2-ylcarbamoyl) sulfamoyl]-1-methylpyrazole-4-carboxylate	1989
Dithiopyr	Dimethyl 2-(difluoromethyl)-4-(2-methylpropyl)-6-(trifluoromethyl) pyridine-3,5-dicarbothioate	1991
Halosulfuron-methyl	Methyl 3-chloro-5-(4, 6-diamethoxypyrimidin-2-yl) carbamoyl)-1-methylpyrazole-4-carbosylate	1991
Sulcotrione	2-(2-chloro-4-mesylbenzoyl) cyclohexane-1,3-dione	1991
Nicosulfuron	2-[(4,6-dimethoxypyrimidin-2-ylcarbamoyl) sulfamoyl]-N, N-dimethyl nicotinamide	1992
Haloxyfop	(RS)-2-{4-[3-Chloro-5-(trifluoromethyl)-2-pyridyloxy] phenoxy} propionic acid	1993
Flumioxazin	N-(7-Fluoro-3,4-dihydro-3-oxo-4-prop-2-ynyl-2H-1,4-benzoxazin-6-phenyl) cyclohex-1-ene-1,2-dicarboxamide	1994
Imazasulfuron	1-(2-Chloroimidazo[1,2-a] pyridin-3-ylsulfonyl)-3-(4,6-dimethoxy pyrimidin-2-pheyl) urea	1994
Methylarsonic acid	Methylarsonic acid	1994
Azimsulfuron	1-(4,6-Dimethoxypyrimidin-2-yl)-3-[1-methyl-4-(2-methyl2H-tetrazol-5-yl) pyrazol-5-sulfonyl] urea	1995

Herbicide	Chemical Nomenclature	Year
Imazapic	2-[(RS)-4-Isopropyl-4-methyl-5-oxo-2-imidazolin-2-yl]-5-methylnicotinic acid	1996
Bispyribac-sodium	Sodium 2,6-bis(4,6-dimethoxypyrimidin-2-yloxy) benzoate	1998
Iodosulfuron-methyl sodium	Sodium ({[5-iodo-2-(methoxycarbonyl) phenyl] sulfonyl} carbamoyl) (4-methoxy-6-methyl-1,3,5-triazin-2-yl) azanide	2000
Imazamox	2-[(RS)-4-Isopropyl-4-methyl-5-oxo-2-imidazolin-2-yl]-5-methoxymethyl nicotinic acid	2001
Mesotrione	2-(4-mesyl-2-nitrobenzoyl) cyclohexane-1,3-dione	2001
Mesosulfuronmethyl	Methyl 2-[(4,6-dimethoxypyrimidin-2-ylcarbamoyl) sulfamoyl]-a-(methane sulfonamido)-p-toluate	2001
Topramezone	[3-(4,5-Dihydro-1,2-oxazol-3-yl)-4-mesyl-o-tolyl] (5-hydroxy1-methylpyrazol-4-yl) methanone	2006
Orthosulfamuron	1-(4,6-Dimethoxypyrimidin-2-yl)-3-[2-(dimethylcarbamoyl) phenylsulfamoyl] urea	2007
Tembotrione	2-[2-Chloro-4-(methylsulfonyl)-3-[(2,2,2-(trifluoroethoxy) methyl] benzoyl]-1,3-cyclohexanedione	2007
Metazosulfuron	1-{3-Chloro-1-methyl-4-[(5RS)-5,6-dihydro-5-methyl-1,4,2-dioxazin-3-] pyrazol-5-sulfonyl}-3-(4,6-dimethoxypyrimidin-2-pheyl) urea	2011
Monosulfuron	2-[(4-Methylpyrimidin-2-yl) carbamoylsulfamoyl] benzoic acid	2012

Sources: The Pesticide Manual (BCPC), 1991 and 2012, and Pesticide Properties Database, University of Hertfordshire, 2015.

Herbicide

It is a chemical used to kill so targeted plants.

In 1944 - discovery of 2.4-D Na salt as a land mark in herbicide usage.

Classification of Herbicides

I. Based on Method of Application

Soil applied herbicides: Herbicide act through root and other underground parts of weeds.

e.g. Fluchloralin.

Foliage applied herbicides: Herbicide primarily active on the plant foliage *e.g.* **Glyphosate, Paraquat**

II. Based on Mode of Action

Selective herbicide: An herbicide is considered as selective when in a mixed growth of plant species, it kills so species without injuring the others. *e.g.* **Atrazine**

Non-selective herbicide: It destroys majority of treated vegetation. *e.g.* **Glyphosate, Paraquat**

III. Based on mobility

Contact herbicide: A contact herbicide kills those plant parts with in direct contact *e.g.* Paraquat

Translocated herbicide: Herbicide which tends to move from treated part to untreated areas through xylem/phloem depending on the nature of its molecule. *e.g.* **Glyphosate.**

IV. Based on Time of Application

Plant application: Application of herbicide before sowing or along with sowing. Either it is foliar applied or incorporated in soil soon after its application. Pre-plant foliar spraying of glyphosate to control perennial weeds like *Cyperus rotundus* and pre-plant soil incorporation of Fluchloralin to control weeds in groundnut

Pre-emergence: Herbicide is applied to soil soon after sowing a crop before emergence of weeds *e.g.* Atrazine, Pendimethalin, Butachlor, Thiobencarb, Pretilachlor.

Post-emergence: When herbicide is applied to killyoung weeds standing in the crop plants or application after the emergence of weed and crop. *e.g.* Glyphosate, Paraquat, 2,4-D Na Salt.

Early post emergence: Application of herbicide in the slow growing crops like potato, sugarcane, 2-3 week after sowing is classified as early post emergence.

Classification of Herbicides Based on Chemistry

Sl.No.	*Chemical Group*	*Examples*
1.	Aliphatics	Allyl alcohol, TCA, Dalapon
2.	Amides and acetamides	Alachlor, Butachlor, Propanil, Solan
3.	Arsenicals	AMA, MSMA, Cacodylic acid
4.	Benzoics and phenyl acetates	Bifenox, Dicamba, Chloramben
5.	Bipyridiliums	Paraquat, Diquat, Cyperquat

Sl.No.	Chemical Group	Examples
6.	Carbomates	Asulam, Barban
7.	Dinitroanilines	Fluchloralin, Isoproturan, Nitrofluralin, Pendimethalin
8.	Diphenyl ethers	Nitrofen, Nitrofluorfen, oxyfluorfen
9.	Nitriles	Bromoxynil, Dichlobenil
10.	Phenols	PCP, Dinoseb
11.	Phenoxy acids	2,4-D, 2,4-DB, MCPA, MCPB, 2,4,5-T
12.	Pyridazinones	Norflurazon, Oxapyrazon, Pyrazon
13.	Thiocarbomates	Benthiocarb, Diallete, Molinate, Triallete, EPTC
14.	Triazines	Atrazine, Atryn, tribuzin, Propazine, Simazine
15.	Triazoles	Amitrole, Amitrole-T
16.	Uracils	Bromacil, Isocil, Terbacil
17.	Ureas	Monouron, Diuron, Cycluron, Fenuron, Linuron
18.	Unclassified	DCPA, thazole, Cypromid, Glyphosate, Picloram

Source: TNAU portal.

Recommended Herbicides for different Crops in India

Crop	Herbicide	Dose (kg ai/ ha)	Trade Name and Formulation	Time of Application
Rice	Butachlor	1.25	Machete 50 per cent EC Delchlor 50 per cent EC	Pre-emergence
	Thiobencarb	1.25	Thunder 50 per cent EC Saturn 50 per cent EC	Pre-emergence
	Anilophos	0.40	Arozin 30 per cent EC Aniloguard 30 per cent EC	Pre-emergence
	Fluchloralin	0.90	Basalin 45 per cent EC	Pre-emergence
	Pretilachlor	0.50	Refit 50 per cent EC	Pre-emergence
	Pendimethalin	0.90	Stomp extra 38.6 per cent EC	Pre-emergence
	2,4-D Na salt	1.00	Fernoxone 80 per cent SS	Post-emergence

Crop	Herbicide	Dose (kg ai/ ha)	Trade Name and Formulation	Time of Application
Rice (Upland direct sown)	Thiobencarb	1.25	Saturn 50 per cent EC	Pre-emergence (8 DAS)
	Pretilachlor	0.45	Refit 50 per cent EC	Pre-emergence
Sorghum	Atrazine	0.25	Atrataf 50 per cent WDP	Pre-emergence
Ragi (Trans-planted)	Butachlor	1.25	Machete 50 per cent EC	Pre-emergence
	Pendimethalin	0.75	Stomp 30 per cent EC	Pre-emergence
Maize	Atrazine	0.25	Atrataf 50 per cent WDP	Pre-emergence
Cumbu	Atrazine	0.25	Atrataf 50 per cent WDP	Pre-emergence
Cotton	Alachlor	1.00	Dual 50 per cent EC	Pre-emergence
	Fluchloralin	1.00	Basalin 45 per cent EC	Pre-emergence
	Pendimethalin	1.00	Stomp extra 38.6 per cent EC	Pre-emergence
	Diuron	0.40	Karx 50 per cent WP	Pre-emergence
Ground-nut	Alachlor	1.00	Dual 50 per cent EC	Pre-emergence
	Fluchloralin	0.90	Basalin 45 per cent EC	Pre-emergence
Sunflower	Fluchloralin	0.90	Basalin 45 per cent EC	Pre-emergence
	Pendimethalin	0.90	Stomp extra 38.6 per cent EC	Pre-emergence
Vege-tables	Fluchloralin	1.00	Basalin 45 per cent EC	Pre-emergence
	Pendimethalin	1.00	Stomp 30 per cent EC	Pre-emergence
Sugarcane	Atrazine	1.00	Atrataf 50 per cent WDP	Pre-emergence
Pulses	Fluchloralin	0.70	Basalin 45 per cent EC	Pre-emergence
	Pendimethalin	0.60	Stomp extra 38.6 per cent EC	Pre-emergence
Wheat	Isoproturon	0.60	Arelon 75 per cent WP	Pre-emergence
Cropping Systems				
Sorghum + Cowpea	Pendimethalin	0.90	Stomp extra 38.6 per cent EC	Pre-emergence
Sugarcane + Pulses	Thiobencarb	1.25	Saturn 50 per cent EC	Pre-emergence
Maize + Soybean	Pendimethalin	1.00	Stomp extra 38.6 per cent EC	Pre-emergence

Herbicides Used for Aquatic Weed Control

Herbicide	*Types of Weeds*	*Rate*
Sodium arsenite	Submerged weeds	5-8 ml L^{-1}
Copper sulphate	Submerged weeds + algae	0.5-2.0 mg L^{-1}
Hydrogen peroxide	Submerged weeds	10-20 mg L^{-1}
Dalapon	Emergent grass weeds	18-25 kg^{-1}
2,4-D	Submerged weeds, Free floating and emergent weeds	1.0 mg L^{-1}
Diuron	Algae; submerged floating and emerged weeds	0.5 to 1.5 mg L^{-1}
Triazines	Algae; submerged and free-floating weeds	0.5 to 1.0 mg L^{-1}
Paraquat	Submerged and'floating weeds	0.5 mg L^{-1}
Diquat	Floating and emerged weeds	1.0 kg^{-1}
Glyphosate	Emergent and floating weeds	1.8-2.1 kg^{-1}

Herbicide Label Information

Several symbols are printed on the label. There are four categories of toxicity and each of which is indicated by an appropriate symbol as under:

Sl. No.	*Toxicity Category*	*Acute oral Toxicity-LD_{50} (mg/kg)*	*Colour of Triangle*	*Signal Word Required on Label*	*Warning Symbol on the Label*
1.	Extremely toxic	0 – 50	Bright red	**POISON** printed in Red	Skull and cross bones
2.	Highly toxic	51 - 500	Bright yellow	**POISON** printed in Red	-
3.	Moderately toxic	501 - 5000	Bright blue	DANGER	-
4.	Slightly toxic	5000	Bright green	CAUTION	

Herbicide Application Equipment's and Calibration of Sprayers

Herbicides are applied in the form of solution or granules. Solution form of herbicides are to be applied as spray. Spraying is the most common method of applying solutions of herbicides.

Granules are applied by hand or with the help of a granular applicator.

Herbicide Spray

Spray is defined as liquid discharged into particles and scattered/dispersed in the form of minute droplets.

Sprayer

It is an equipment used for spraying chemicals.

Types of Nozzles

a) **Cone type:** Cone nozzles are widely used for **insecticide and fungicide** applications as also for spot application of post-emergence herbicides. There are two types of cone nozzles.
   i) Solid cone: The liquid passes centrally through the nozzle to fill the air cone.
   ii) Hollow cone: The liquid emerges from the orifice as a hollow cone.

b) **Fan nozzle:** Ideal for spraying flat surfaces and for band spraying in row crops.

c) **Flooding or Deflector type nozzle:** produces a wide-angle fan pattern of spray.

Herbicide spraying is mostly done with fan and flooding or deflector type nozzles as they produce a more even distribution of spray and uniform coverage than the cone spray types.

Calculating Commercial Formulation

The commercial formulation of herbicide required can be worked out using the formula

$$\text{Formulated product required} = \frac{\text{Role of a.i./ha to be applied}}{\text{Active ingredient (per cent)}} \times 100$$

For example, if one needs to apply Diuron @ 2 kg a.i./ha and the commercial formulation contains 80 per cent a.i., the total product required per ha would be:

$$= \frac{2 \times 100}{80} = 2.5 \text{ kg}$$

Herbicides Mode Action

1) 2, 4-D (2,4-Dichloropnenoxy Acetic Acid)

It is selective translocated herbicide and most widely used to control dicot weeds (Broad Leaf Weeds) in cereals and sugarcane as post emergence application.

It is cheap, easily available, easy to apply and non-poisonous to human beings and animals. It is available in three formulations as Sodium Salt (Fernoxone 80 wp or Agrosodium), Dimethylamine (Agrodor 96 or Weedar 96) and Ethylester (Knockweed 36 EC or Knock weed 4G). Sodium salt is white powder with characteristics odour, low volatile and is soluble in water. The dimethylamine is light brown liquid forms emulsion in water and is non-volatile. Ethylester is light yellow liquid may be low or high volatile and forms milky emulsion with water and it is absorbed by weeds at faster rate than other forms so more effective for control of weeds. It is applied @ 1 to 1.5 kg a.i/ha in 500 to 600 liters as post emergence spray.

2) 2,4,5-T: (2,4, 5 – Trichlorophenoxy Acetic Acid) and 2,4,5-TP (2,4,5-Trichloropropionic Acid):

Both the herbicides are similar to 2,4-D in properties and mode of action and useful for controlling bushes and woody weeds.

3) Dicamba: (3,6 –Dichloro-2-Methoxybenzoic Acid)

It is selective translocated herbicides. Useful to control broad leaf weeds and used @ 0.5 to 3 kg a.i/ha as postemergence spray.

4) Simazine: (2, Chloro-4,6 –bi(Ethylamino) s-triazine)

Atrazine (2-Chloro-4- (ethylamino -6- isoprophylamino-5-triazine. Both the herbicides are selective, translocated and as pre-emergence spray for killing broad leaf weeds and grasses in sugarcane, maize, jowar, potato, grapes *etc.* The selective rates are 0.5 to 2.5 kg a.i/ha in 500 to 600 liters of water. Atrazine is more soluble in water than simazine. Therefore, atrazine is more useful in dry farming areas.

5) Mertibuzine (4-Amino -6-tert-butyl- 3-(Methylthio) as triazine -5(4H)-one)

It is broad spectrum selective herbicides with both, soil and shoot activity and found useful in sugarcane, potato, tomato, soybean, *etc.* The effective rates are 0.2 to 1 kg a.i/ha.

6) Paraquat (1,1 –diethyl -4- bipyridinium Ion and Diquat (6,7 –dihydrodipyrido (1,2:2, I-C) Pyrazinediiumaion)

Both the herbicides are contact, non-selective with zero persistence in the soil. Useful in sugarcane before germination and as direct spray after germination in sugarcane and orchards. Also useful as fallow application before sowing and also in non-cropped areas. 1 to 2 kg a.i/ha as post emergence spray.

Paraquat is availability trade name gramoxone as 24 SL and diquat as reglone 20 EC.

7) Benthiocarb or Thiobencarb: (S-(4-Chlorobenzyl) N, N – Diethyl – Thicarbamate)

It is effective for control of *Echinochloa* spp. in rice. It is used as pre-emergence spray @ 1 to 2 kg a.i/ha.

8) Alachlor (92-Chloro – 2,6- Diethyl- N- (Methoxymethyl) Acetanilide)

It is very annual grasses and certain broad leaf weeds in groundnut, soybean, cotton potatoes, maize, sugarcane and vegetables. It is applied @ 1 to 2 kg a.i/ha.

9) Butachlor N (Butoxymethy 2,6 –diethylacetanilide)

It is selective pre-emergence herbicides used in seedbed and transplanted rice for control of *Echinochloa* spp. and most other annual weeds. The common rates are 1 to 2 kg a.i/ha. Upto 3 to 5 days after sowing in seeded rice and upto 10 days after transplanting in transplanted rice. It is available by trade name Machete as 50 EC or 5 per cent granules.

10) Fluchloralin: (N-(2-Chloroethyl)-2,6-dinitro-N-prophyl) – 4 (trifluromethyl) amine)

It is selective pre-emergence herbicide effective against large number of annual weeds in cotton, soybean, jute, chickpea, sunflower, onion and certain solanaceous vegetables. Pre-planting soil incorporation of the herbicide give better results as it is volatile. It is applied @ 1 to 1.5 kg a.i/ha.

11) Pendimethalin: (N-(1-ethylprophyl) - 3, 4- dimethyl -2,6-dinitrobenzenamine)

It is primarily selective pre plant, soil incorporated herbicide found effective against number of annual weeds particularly grasses in cotton, soybean, groundnut and pea. It is also used as pre-emrgence herbicide in maize, rice, small grains, onion and potato. The selective rates are 1 to 1.5 kg a.i/ha.

12) Glyphosate (N- (Phosphomethyl) (glycine)

It is post emergence, non-selective, translocated herbicide. It is highly shoot mobile and weak residue herbicide and can be used against perennial weeds few days before sowing or as post emergence spray in non-cropped areas. Very effective to control perennial weeds like nutgrass. Recommended dose is 1 to 3 kg a.i/ha as post emergence spray when weeds are in active stage of growth. It is useful for killing perennial weeds effectively in fruit crops and sugarcane as directed spray.

Approved Uses of Registered Herbicide Combinations

Herbicide Name & Approved Crops	*Weed Species*	*Dosage /ha*		*Dilution in Water (Litres)*	*Waiting Period/ PHI between Last Application & Harvest (Days)*
		a.i. (gm/ kg)	*Formulation in (gm/ ml / kg/ ltr)*		
Anilofos 24% +2,4-D ethyl Ester 32% EC					
Transplanted rice	*Echinochloa crusgalli, Echinochloa colonum, Ischaemum rugosum, Fimbristylis miliacea*	(0.24+ 0.32) to (0.36 + 0.48)	1-1.5 ltrs.	300	90
Bensulfuron methyl 0.6%+Pretilachlor 6% GR					
Transplanted Rice	*Echinochloa crusgalli, Echinochloa colonum, Cynodon dactylon Cyperus iria, Cyperus difformis, Cyperus rotundus, Fimbristylis miliacea, Ludwigia parviflora, Marselia quadrifolia, Enhydra fluctuans, Sphenoclea zeylanica, Eclipta alba, Ammania baccifera*	60 + 600	10 kg	N.A.	88
Carfentrazone ethyl 0.43% + Glyphosate 30.82% EW					
Tea	*Ageratum conyzoides, Bidenspilosa Borreria* sp., *Crassocephalumcr epidioides, Cynadon* sp., *Cyperous* sp., *Digitaria* sp., *Eleusine indica, Mimosa* sp., *Mltracarpus villosus, Oxalis sp.*	12.90 + 924.60	3000	500	7

Herbicide Name & Approved Crops	Weed Species	Dosage /ha		Dilution in Water (Litres)	Waiting Period/ PHI between Last Application & Harvest (Days)
		a.i. (gm/ kg)	Formulation in (gm/ ml / kg/ ltr)		
Non-cropped area	*Ageratum conyzoides, Axonopus* sp., *Brachiaria* sp., *Commelina* sp., *Cynodon dactylon, Cyperous* sp., *Digitaria* sp., *Eleusine indica, Imperata cylendrica, Lantana camera, Parthenium sp.*	12.90 +924.60	3000	500	--
Carfentrazone ethyl 20% + Sulfosulfuron 25% WG					
Wheat	*Phalaris minor, Avena ludoviciana, Chenopodium album, Medilotus alba, Rumex* spp.	20+25 +750 ml Surfactant	100	300	110
Clodinafop Propargyl 15% + Metsulfuron Methyl 1% WP					
Wheat	*Phalaris minor, Avena fatua, Chenopodium album, Melilotus* sp., *Fumaria parviflora, Vicia sativa, Rumex* sp., *Anagallis arvensis, Coronopus didymus, Lathyrus* sp., *Convolvulus arvensis*	60+4	400	375 (Add 1250 ml surfactant at the time of sparying)	100
Clodinafop propargyl 9% + Metribuzin 20% WP (W/W)					
Wheat	*Phalaris minor, Chenopodium album, Melilotus* sp., *Vicia sativa, Rumex* sp., *Medicago* sp., *Cronopus didymus, Dinebra vetroflexa*	54+120	600	300	120

Herbicide Name & Approved Crops	Weed Species	Dosage /ha		Dilution in Water (Litres)	Waiting Period/ PHI between Last Application & Harvest (Days)
		a.i. (gm/ kg)	Formulation in (gm/ ml / kg/ ltr)		
Clomazone 20%+2,4-D EE 30% EC					
Transplanted Rice	*Echinochloa colonum, Echinochloa crusgalli, Cyperus iria, Cyperus difformis, Eclipta alba, Leptochloa chinensis, Panicum repens, Fimbristylis miliacea, Marsilea quadrifoliata Ludwigia parviflora.*	0.250-0.375	1.25 ltrs.	500	100-110
Fenoxaprop-p-ethyl 7.77% w/w + Metribuzin 13.6% w/w EC					
Wheat	*Phalaris minor* (Little seed canary grass), *Chenopodium album* (Lambs quarter), *Lathyrus aphaca* (Meadow Pea), *Rumes sp.* (Golden dock), *Melilotus spp.* (Sweet clover), *Avena ludoviciana*	100+175	1250	375	110
Fluazifop-p-butyl 11.1% w/w + Fomesafen 11.1% w/w SL					
Soybean	*Echinochloa colona, Digitaria* sp., *Eleusine indica, Dactyloctenium aegyptium, Brachiaria reptans, Commelina benghalensis, Digera arvensis, Trianthema* sp., *Phyllanthus niruri, Aclypha indica, Dinebra arbica*	250	1000	500	71

Herbicide Name & Approved Crops	Weed Species	Dosage /ha		Dilution in Water (Litres)	Waiting Period/ PHI between Last Application & Harvest (Days)
		a.i. (gm/ kg)	Formulation in (gm/ ml / kg/ ltr)		
Groundnut	*Echinochloa colona, Digitaria* sp., *Eleusine indica, Dactyloctenium aegyptium, Commelina benghalensis, Eluropus villosus, Indigofera glandulosa, Chloris barbata, Trianthema* sp., *Digera arvensis, Cleome viscose, Phyllanthus niruri, Amaranthus virdis, Cyperus sp.*	250	1000	500	82
Hexazinone 13.2% + Diuron 46.8 % WP					
Sugarcane	*Enchinochloa colonum, Dactylotenium aegyptium, Trianthema monogyna, Amaranthus virdis, Ipomea* spp., *Cyperus rotundus, Cyperus esculentus, Setaria* spp., *Parthenium hysterophorus, Euphorbia hirta*	1200 gm (264+936)	2 kg	500	282-306
Indaziflam 1.65% w/w (2%w/v) + Glyphosate Isopropylammonium 44.63% w/w (40%w/v) SC					
Tea	*Ageratum* sp., *Borreria* sp., *Eleusine indica*	50+1000 to 70 + 1400 g.a.i/ha	2500 to 3500 ml/ha	500 L	14 days
Imazethapyr 35% + Imazamox 35% WG					
Soybean	*Echinochloa colonum, Dinebra arabica, Digitaria sanguinalis, Brachiaria mutica Commelina benghalensis, Euphorbia hirta*	70 g a.i/ha + MSO adjuvant @ 2 ml/l of water	100 g MSO Aadjuvant @ 2ml/l of water	375-500	56

Herbicide Name & Approved Crops	Weed Species	Dosage /ha		Dilution in Water (Litres)	Waiting Period/ PHI between Last Application & Harvest (Days)
		a.i. (gm/ kg)	Formulation in (gm/ ml / kg/ ltr)		
Groundnut	*Echinochloa Colonum, Digira arvensis, Commelina benghalensis, Euphorbia hirta, Amaranthus viridis, Physalis minima*	70 g a.i/ha + MSO adjuvant @ 2 ml/l of water	100 g MSO Aadjuvant @ 2ml/l of water	375-500	83
Cluster bean	*Echinochloa colonum, Euphorbia* spp., *Digitaria arvensis, Amaranthus viridis*	70 g a.i./ha + MSO adjuvant @ 2m/l of water	100 g/ha + MSO adjuvant @ 2m/l of water	500	64
Red gram	*Euphorbia* spp., *Amaranthus viridis*	70 g a.i./ha + MSO adjuvant @ 2m/l of water	100 g/ha + MSO adjuvant @ 2m/l of water	375-500	125
Mesoulfuron Methyl 3% + Iodosulfuron Methyl Sodium 0.6% WG					
Wheat	*Phalaris minor, Medicago denticulata, Chenopodium album, Melilotus* sp., *Rumex* sp., *Anagallis arvensis, Coronopus didymus, Lathyrus aphaca, Fumaria parviflora*	(12+2.4)	400 ml.	400-500 + Surfactant (Genopol LRO fluid) @ 500 ml/ ha	96

Herbicide Name & Approved Crops	***Weed Species***	***Dosage /ha***		***Dilution in Water (Litres)***	***Waiting Period/ PHI between Last Application & Harvest (Days)***
		a.i. (gm/ kg)	***Formulation in (gm/ ml / kg/ ltr)***		
Metsulfuron Methyl 10% + Chlorimuron ethyl 10% WP					
Transplanted Rice (Pre-emergence application-3 DAT	*Cyperus iria, Cyperus difformis, Fimbristlylis miliaceae, Eclipta alba, Ludwigia parviflora, Cyanotis axillaries, Monocoria vaginalis, Marsilea quadrifoliate*	4	20 gm.	300	90
Metsulfuron Methyl 10% + Carfentrazone ethyl 40% DF					
Wheat	*Rumex dentatus, Rumex spinosus, Medicago denticulate, Malva parviflora, Lathyrus aphaca, Chenopodium album, Melilotus alba, Melilotus indica, Anagallis arvensis, Solanum nigrum, Vicia sativa, Convolvulus arvensis*	25	50	300	100
Oxyflurofen 2.5% + Glyphosate (Isopropyl amime salt)41% SC(w/w)					
Tea	*Ageratrum conyzoids, Cyperous sp., Borreria bispida, Pospalumcon jugatum, Digitaria ciliaris*	50+820	2000	500L/ha.	14
Pendimethalin 30%+ Imazethapyr 2% EC					
Soybean	*Echinocloa crusgalli, Digera arvensis, Commelina benghalensis, Amaranthus viridis, Portulaca oleracea*	(750+50) to (900+60)	2.5-3.0 ltrs	500-600	90

Herbicide Name & Approved Crops	Weed Species	Dosage /ha		Dilution in Water (Litres)	Waiting Period/ PHI between Last Application & Harvest (Days)
		a.i. (gm/ kg)	Formulation in (gm/ ml / kg/ ltr)		
Penoxsulam 0.97% w/w + Butachlor 38.8% w/w SE					
Transplanted Rice	*Echinochloa colonum, Echinochloa crusgalli, Cyperus iria, Cyperus difformis, Marsilia quadrifoliata, Alternanthera* spp.	820 g.a.i/ha	2000ml/ha	750	60
Penoxsulam 1.02 % + Cyhalofop-butyl 5.1% OD					
Rice (Direct seeded Rice)	*Echinochloa colona, Echinochloa crusgalli, Leptochloa chinesis, Eleusine indica, Alternanthera sessilis, Caesulia axillaris, Cyperus* spp.	120-135	2000-2250	300-500	60
Rice (Trans-planted Rice)	*Echinochloa colona, Echinochloa crusgalli, Leptochloa chinesis, Caesulia axillaris, Cyperus difformis, Cyperus* spp.	120-135	2000-2250	300-500	60
Pretilachlor 6% + Pyrazosulfuron Ethyl 0.15%(H)					
Paddy	Grassy weeds, Broad Leave, Sedges	600+15	10	-	83
Pretilachlor 6.0% + Pyrazosulfuron Ethyl 0.15% GR					
Transplanted Paddy	*Echinochloa Colonum Echinochloa Crusagalli Ludwigia paviflora, Elipta alba, Leptochola chinensis, Monochoria vaginalis, Cyperus difformis, Cyperus iria, Fimbristylis miliaceae*	600	10	-	83
Pretilachlor 30.0% + Pyrazosulfuron Ethyl 0.75% WG					

Herbicide Name & Approved Crops	Weed Species	Dosage /ha		Dilution in Water (Litres)	Waiting Period/ PHI between Last Application & Harvest (Days)
		a.i. (gm/ kg)	Formulation in (gm/ ml / kg/ ltr)		
Paddy (Direct seeded Rice)	*Echinochloa colona, Eleusine indica, Leptochloa chinensis, Abernanthe sessllis, Parthenium hysterophrous, Cyprus iria, Cyprus difformis, Fimbristylis miliacea*	15+600	2000	375	113
Propaquizafop 2.5% + Imazethapyer 3.75% w/w ME					
Soybean	**Grassy weeds**: *Dactyloctnium aegyptiumm, Echinocloa colonum Eleusine indica, Digitaria sanguinalis* **BLW:** *Commelina benghalensis, Euphorbia hirta, Digera arvensis, Amaranthus viridis*	50+75	2000	500	80
Propaquizafop 5% + Oxyfluorfen 12% w/w EC					
Onion	**Grassy weeds:** *Digitaria sanguinalis, Dactyloctnium aegyptium, Echinocloa colona, Eleusine indica* **BLW:** *Chenopodium album, Digera arvensis, Amaranthus viridis*	43.75 +105	875	500	7
Pyrithiobac Sodium 6% EC w/w+ Quizalofop-ethyl 4% EC w/w MEC					

Herbicide Name & Approved Crops	***Weed Species***	***Dosage /ha***		***Dilution in Water (Litres)***	***Waiting Period/ PHI between Last Application & Harvest (Days)***
		a.i. (gm/ kg)	***Formulation in (gm/ ml / kg/ ltr)***		
Cotton	*Trianthema* spp., *Digera* spp., *Celosia argentia, Dinebra retroflexa, Digitaria marginate*	(60+40) to (75+50) g.a.i/ ha	1.0-1.25 Ltr/ ha	500	160
Sulfentrazone 28% + Clomazone 30% WP					
Soybean	*Commelina benghalinsis, Acalypha indica, Corchorus* spp., *Euphorbia* spp., *Parthenium hysterophorus, Echinochloa colonum, Brachiaria* spp., *Dinebra* spp., *Cyperus rotundus*	350+375	1250	500	84
Sugarcane	*Amaranthus viridis, Trianthema* spp., *Digera arvensis, Physalis* spp., *Brachiaria* spp., *Cynodon dactylon, Echinochloa* spp., *Dactyloctenium aegtptium, Cyperus rotundus*	700+750	2500	500	302
Sulfosulfuran 75%+ Metsulfuron Methyl 5%WG					
Wheat	*Phalaris minor, Chenopodium* sp., *Medicago denticulata, Coronopos dedymus, Rumex* spp., *Melilotus alba, Anagallis arvensis*	30+2	40 gm	250-500 + surfactant 1250 ml/ ha	110
Sodium Aceflourofen 16.5% + Clodinafop Propargyl 8% EC					

Herbicide Name & Approved Crops	**Weed Species**	**Dosage /ha**		**Dilution in Water (Litres)**	**Waiting Period/ PHI between Last Application & Harvest (Days)**
		***a.i.* (gm/ kg)**	**Formulation in (gm/ ml / kg/ ltr)**		
Soybean	*Acalypha indica, Alternanthera philoxeroides, Amaranthus* spp., *Celosia argentea, Cleome viscose, Commelina benghalinsis, Dactyloctanium, Digera arvensis, Digitaria sanguinalis,* Echinochloa spp., *Eleusine indica, Euphorbia* spp., *Parthenium* spp., *Phyllanthus niruri, Physalis minima, Stellaria media, Trianthema monogyna*	80 + 165	1000	500	61
Triafamone 20% + Ethoxysulfuron l0% WG					
Rice Transplanted	*Echinochloa colona, Echinochloa crusgalli, Cyperus rotundus, Cyperus difformis, Fimbistylis miliaceae, Marsilea quadrifolia*	44+22.5	225	300	83
Rice Direct Seeded	*Echinochloa colona, Cyperus rotundus, Digera arvensis, Commelina benghalinsis.*	44+22.5	225	300	83

Source: Directorate of Plant Protection, Quarantine & Storage, Central Insecticides Board & Registration Committee, Faridabad, 2020.

India has more than 800 pesticide formulators. Herbicides under 9(4) registration are formulated and marketed by many companies. A few hundreds of formulations of registered herbicides are available in Indian market. Some major formulations with brand names are given below.

Brand Names and Manufacturers of Herbicides

Herbicide	*Brand Names and Manufacturers*
2,4-Dichlorophenoxy acetic acid	80% WP (sodium salt): Fernoxone (Syngenta), Weedmar-80 (Dhanuka Agritech Ltd.), Chem-D (Shaw Wallace), K-D (Krishi Rasayan), TCM 2,4-D sodium salt (TCM), Safaya (Devidayal), Weed Killer (Okruti), Volvo (UPL Ltd.), Heera super (Hindustan Pulverising), Kaal (Plant Remedies), Cut Out (Crystal), Naagsol (Multiplex), Super hit (Insecticide India), 2,4-D Agan (ADAMA)
	58% SL (amine salt): Champion (Herbicide India), Chem-amine (Shaw Wallace), Naagmin (Multiplex), Heera (Hindustan Pulverising), Amine ester (Dhanuka Agritech Ltd.), Shaktiman amine eseter (Indo Gulf), Weed out (Bharat Rasayan), PC 2,4-D (Punjab Chemicals), Aminosol (Solas Crop Tech), Twister (Insecticide India), 2,4-D Main (ADAMA), Weedcel Super (Excel)
	38% EC (ethyl ester): Slash (Monsanto), Weedmar (Dhanuka Agritech Ltd.), Agni (FIlnd), Chem-est (Shaw Wallace), Smash (Nagarjuna), Heera 44 (Hindustan Pulverising), Anu weed (Anu), Champion (Herbicide India), Weed burn (Dara Chem), Nagester (Multiplex), Goldiseed (Swaraj), Electron (Devidayal), Super star (JU Pesti), Plant 2,4-D (Plant Remedies),
	Eleminator (Hem-an), Tusker (Shri Ram Agro), Cut out and Cut Short (Crystal), Kill Herb (Tropical Agro), Hit (Insecticide India)
	34% EC (ethyl ester): K-D (Krishi Rasayan)
	20% WP (ethyl ester): Anuweed (Anu), Kill herb (Tropical Agro), Champion (Herbicide India)
	4.5% Gr (ethyl ester): Smash (Nagarjuna), Cut Out (Crystal)
Alachlor	50% EC: Lasso (Monsanto), Alachlor (Dhanuka Agritech Ltd.), Attack
	25%EC: Catch (Isagro) (Nagarjuna), Allert (Tropical Agro), Dorado (PI Ind.);
	10% Gr : Lasso (Monsanto), Attack (Nagarjuna)

Herbicide	Brand Names and Manufacturers
Anilophos	30% EC: Arozine (Bayer), Sumo (DuPont), Anilogourd (Gharda), Hilanil (HIL), Anilodhan (Dhanuka Agritech Ltd.), Surya (EID Parry), Anilohit (Hindustan Pulverising), Ricil (Dow), Anilostar (Shaw Wallace) Anilotaf (Rallis), Anugaurd (Anu), Anilophos (Nagarjuna), Anilex (FIL Ind.), Avail (Herbicide India), Tag guard (Tropical Agro), Anilon (Sulfur Mills), Anilocide and Quatar (Bharat Rasayan), Anilospeak (TAC), Anilovip (Godrej), Anilocide and Army (Insecticide India), Rainbow (Gujarat Krishi), Shreeramanilo (Shri Ram Chem), Strong (Dara Chem), Libra (PI Ind.), Angola (JU Pesticide), Hemilophos (Hem-an), Foster (Devidayal), Anilon and Aniloshree (Crystal), Nidan (Gujarat Insecticides Ltd.)
	18% EC: Rico (Bayer)
	60% LC: Bulk (Bayer)
	2% Gr: Arozine (Bayer)
Atrazine	50% WP: Solaro (PI Ind.), Atrataf (Rallis), Dhanuzine (Dhanuka Agritech Ltd.), Gesaprim (Syngenta), Atrafil (Indofil), Anutaf (Anu), Atrahit (Hindustan Pulverising), Rasayanzin (Krishi Rasayan), Surya (Nagarjun), Atravip (Godrej), Atrasul (Sulfur Mills), Tagtaf (Tropical Agro), Atrazine (EID Parry), Attack (Devi Dayal), Pearl (UPL Ltd.), Ultra (JU Pesti), Nagzine (Multiplex), Merkzine (Merk Fed), Atrazine (Coromondal), Tofa (Bharat Rasayan), SMP (FIL Industries), Atrazide (Hem-an), Sreejon and Reed off (Crystal), Ratrazine (Shri Ram Chem), Petra (Cheminova), Atranex (ADAMA), Atrea (Heranba), Xtrazine (GSP), Atracel (Excel), Citra (Punjab Chemicals), Attrastar (Shaw Wallace)
Azimsulfuron	50% DF: Segment (DuPont)
Bispyribac-sodium	10% SC: Nominee Gold (PL Ind), Adora (Bayer), TKS-PYRI (Tata Chemicals), Taarak (Rallis)
Butachlor	50% EC: Machete and Aristo (Monsanto), Check-out (UPL Ltd.), Trap (Isagro), Bilchlor (Bayer), Teer (Rallis), Dhanuchlor (Dhanuka Agritech Ltd.), Weedkill (Sudershan), Hiltachlor (HIL), Biochlor and Hunter (Hindustan Pulverising), Delchlor (Coromondal), Aimchlor (AIMCO), Rasayanchlor (Krishi Rasayan), Anuchlor (Anu), Weeper (Nagarjun), Estachlor (ECI), Starchlor (Show Wallace), Sterchlor (Stirling), Finish (FIL Ind.), Punch (Herbicide India), Nagchlor (Multiplex),

Herbicide	Brand Names and Manufacturers
	Thunder (Tropical Agro), Weed out (Sulfur Mills), Goldchlor (Swaraj), Butacid and Guru (Bharat Rasayan), Tackchlor (TAC), Gamachlor (Andu), Cenchlor (Century), Butabic (Vikrant), Tarachlor (Tarama), Shaktiman butachlor (Indo-Gulf), Shrirambuta (Shri Ram Chem), Ratcheat (Plant Remedies), Vanchlor (Vantech), Megachlor (Meghmani), Currant (Somanil), Butadel (Del company), Darachlor (Dara Chem), Butavip (Godrej), Betaten
	(Devidayal), Cropchlor ((JU Pest), Merkchlor (Merkfed), Burnchlor (Hem- an), Bumper and Topper (Crystal), Millchlor (Insecticide India), TKS buta (Tata Chem.), Herachlor (Heranba)
	5% Gr: Machete (Monsanto), Trap (Isagro), Dhanuchlor (Dhanuka Agritech Ltd.), Hiltachlor (HIL), Delchlor (Coromandel), Rasayanchlor (Krishi Rasayan), Anuchlor (Anu), Weeper (Nagarjun), Estachlor (ECI), Sterchlor (Stirling), Finish granule (FIL Ind.), Punch (Herbicide India), Thunder (Tropical Agro), Weed out (Sulphur Mills), Butabic-G (Vikrant), Tarachlor (Tarama), Shaktiman butachlor (Indo-Gulf), Ratcheat (Plant Remedies), Vanchlor (Vantech), Butachlor (Dara Chem), Cropchlor ((JU Pest), Bumper (Crystal), Rajchlor (RPC), Millchlor (Insecticide India)
	50% EW: Fastmix (Monsanto), Superfirst (Bharat rasayan), Hunter super (Hindustan Pulverising), Don mix (Dhanuka Agritech Ltd.), Fastest (Anu), Millfast (Insecticide India)
Carfentrazone-ethyl	40% DF : Affinity (FMC India Pvt. Ltd.)
Chlorimuron-ethyl	25% WP: Kloben (DuP ont), Rise (Hin dust an Pulv erising), Curin (Dhanuka Agritech Ltd.), Falone (Crystal), Blumen (ADAMA), Flat (Indofil)
Cinmethylin	10% EC : Argold (BASF)
Clodinafop-propargyl	15% WP : Topic (Syngenta), Dinofop (Dhanuka Agritech Ltd.) Award (Herbicide India), Jhatka (UPL Ltd.), Weeddout (Godrej), Atches (Biostadt), Cloudy (UPL Ltd.), Moolah (Chemtura), Ohm (Punjab Chemicals), Skipper (Coromandel), Rakshak (Cheminova), Topple (Shaw Wallace), Sartaj (Rallis)
Clomazone	50% EC : Command (FMC)

Herbicide	***Brand Names and Manufacturers***
Cyhalofop-butyl	10% EC : Clincher (Dow), Wrap-up (Dhanuka Agritech Ltd.), Tata Cylo (Rallis), Cyhalofop-butyl (Excel), Dentil (Cheminova)
Diclofop-methyl	28% EC : Illoxan (Bayer)
Diuron	80% WP : Klass (Bayer), Diuron (Northern Min.), True (Herbicide India), Naguron (Multiplex), Diurex (ADAMA)
Diclosulam	84% WDG : Strongarm (Dow)
Ethoxysulfuron	15% WG : Sunrice (Bayer)
Fenoxaprop-P-ethyl	10% EC: Puma Power (Bayer), Phenox (Dhanuka Agritech Ltd.) Super Power (Herbicide India), Rider (Crystal), Nagfenop (Multiplex), Ficol (Rallis), Devi Slash (Devidayal)
	9.3% EC: Weepsuper (Bayer)
	6.9% EC: Ricestar (Bayer)
Fluazifop-P-butyl	9% EC: Fusilade Super (Syngenta)
Fluchloralin	45% EC: Basalin (BASF), Dhanulin (Dhanuka Agritech Ltd.), Nagflur (Multiplex), Anulin (Anu), Vicolin (Gujarat Krishi), Herbilin (Herbicide India), Flight (Devidayal)
Glufosinate-ammonium	13.5% SL: Basta (Bayer), Liberty (Bayer)
Glyphosate	41% SL: Roundup (Monsanto), Glycel (Excel), Brake (Biostadt), Sweep (UPL Ltd.), Winner (Cheminova), Noweed (Northern Min.), PI Glypho (PI Ind.), Weedoff (Dow), Krup and Kriup (Krishi Rasayan), Vinash (Sulfur Mills), Catlus (Gujarat Krishi), Groundup (Hindustan Pulverising), Clean up (Indofil), Safal (Tropical Agro), Clear out (Anu), Wonder (Herbicide India), Glycovip (Godrej), Dryphosate (Shaw Wallace), Root Out (Bharat Rasayan), Nag Glypo (Multiplex), Gly-weed (Sabero), Trinnashi (HIL), Irador (Hem-an), Cleanton (Crystal), Touch down (Syngenta), Hijack (Insecticide India), Glycide (Shri Ram Chem), Bonus (JU Pest), Prepare (Bayer), Glyder (Gharda), Leanch (Isagro), Globus (Nagarjuna), Glyphogun (ADAMA), Popular (Aristo), Phosgly (Tata Chem), Glory (Heranba), Run out (GSP), Gladiator (Devidayal), Glycor (Coromandel), Veto (Punjab Chemicals), Glytaf (Rallis), Glyphos (Shri Ram Agro)
	71% SC: Urgent SG (Bayer), Glyphos Ducker (Cheminova), Andhi 71 (Bharat Rasayan), Excel Meera 71 (Excel), Veto Plus (Punjab Chemicals), Killshot (Coromandel), Star71 (Shaw Wallace)

Herbicide	*Brand Names and Manufacturers*
	5% SL: Take-5 (Excel)
	20.2% SL: Gamboy (Excel)
Halosulfuron-methyl	75% SG : Sempra (Dhanuka Agritech Ltd.)
Imazethapyr	10% SL: Persuit (BASF), Inro (PI Ind.), Shikhor (Bharat Rasayan), Cheetah (Hindustan Pulverising), Fervent (Coromandel), Spur (Shaw Wallace), Tata Vaar (Rallis)
	10% EC: Class
Isoproturon	75% WP: Record (DuPont), Bilron (Bayer), Rakshak (Cheminova), Tolcan and Erelon (Bayer), Isogourd (Gharda), Dhanulon (Dhanuka Agritech Ltd.), Isohit (Hindu stan Pu lverising), Nocilon (Dow), Gram inon (Syngenta), Keturon and Krilon (Krishi Rasayan), Anulon (Anu), Norilon (Nagarjuna), Isoper (EID Parry), Isostar (Show Wallace), Miracle (FIL Ind.), Sarpanch (Herbicide India), Isoproturon (Devidayal), Wonder (Tropical Agro), Proton-75 (Sulfur Mills), Goldron (Swaraj), Isolon and Ghashmar (Bharat Rasayan), Isovip (Godrej), Sonaron (Gujarat Krishi), Shree Ram iso (Shree Ram Chem), Shaktiman isoproturon (Indo-Gulf), Passport (Isagro), Delkron (Coromandel), Somilon (Somanil), Kanak (Paushak), Meghalon (Meghm ani), Dararon (Dara Chem), Josh (Sudershan), Eleminron (Hem-an), Sreelon and Kripan (Crystal), Dragon (JU Pesticides), Taurus (PI Ind.), Plant Iso (Plant Remedies), Merklon (Mark Fed), Milron and Skylark (Insecticide India)
	50% WP: Tolcan and Erelon (Bayer), Rakshak (Cheminova), Dhanulon (Dhanuka Agritech Ltd.), Isohit (Hindustan Pulverising), Krilon (Krishi Rasayan), Isostar (Show Wallace), Sarp anch (H erbicid e Ind ia), Isoproturon (Devi Dayal), Wonder (Tropical Agro), Somilon (Somanil), Kanak (Paushak), Proton-50 (Sulfur Mills), Isolon (Bharat Rasayan) Taurus (PI Ind.), Sonaron (Gujarat Krishi), HIL-proturon (HIL)
Linuron	50% WP : Afalon (Aventice), Lorox (Rallis)
Mesosulfuron-methyl + Iodosulfuron-methyl sodium	3.6% WG : Atlantis (Bayer)

Herbicide	*Brand Names and Manufacturers*
Methabenzthiazuron	70% WP : Dhanulin (Dhanuka Agritech Ltd.), Tribunil (Bayer)
Methyl chlorophenoxy acetic acid (MCPA)	40% SL : MCPA amine-salt (Hindustan Pulverising)
Metolachlor	50% EC : Dual (Syngenta)
Metribuzin	70% WP : Sencor (Bayer), Tata Metri (Rallis), Barrier (Dhanuka Agritech Ltd.), Chase (Indifil), Krizine (Krishi Rasayan), Stump (Hindustan Pulverising), Desire (Tropical Agro), Macro (Sulfur Mills), Mohra (Anu), Zero weed (Bharat Rasayan), Jet Strike (Sudershan), Herbi Metri (Herbicide India), JU Merit (JU Pest), Laskar (UPL Ltd.)), Shri Ram metrizine (Shri Ram Chem), Imetar (ECI), Merker (Markfed), Century (FIL Ind), Metshort (Hem-an), Shreecor and Tidemetri (Crystal), Genious (Shri Ram Agro), Hilmetri (HIL), Metribusvip (Godrej), Anchlor (Insecticide India), Metiherb (Cheminova), Metex (Excel), Emax (Nagarjuna), Weedclean (Super Agro), Encor (Devidayal), Metriagan (ADAMA), Grometri (Coromandel), Verdict (GSP)
Metsulfuron-methyl	20% WP: Algrip (DuPont), Metsy (UPL Ltd.), Freedom and Hygrip (Hindstan Pulverising), Bilgrip (Bharat Rasayan), Teg (Rallis), Volt (FIL Ind.), Rado and Algo (Crystal), Hook (Dhanuka Agritech Ltd.), Weed grip (Insecticide India), Metsil-M (Isagro), JU-grip (JU Pesticid e), Dot (Nagarjuna), Nikanor (ADAMA), Metcil (Cheminova), Metro (Heranba), Siege (Punjab Chemicals), Metcil M (Cheminova), Metstar (Shaw Wallace)
Oxadiargyl	80% WP: Topstar (Bayer)
	6% EC: Raft (Bayer)
Oxadiazon	25% EC: Ronstar (Bayer)
Oxyfluorfen	23.5% EC: Goal (Bayer), Oxygold (Indofil), Oxykill (Dhanuka Agritech Ltd.), Crall (Krishi Rasayan), Alto (FIL Ind.), Herbigold (Herbicide India), Waada (Hindustan Pulverising), Honcho (Rallis)
Paraquat dichloride	24% EC: Gramoxone (Syngenta), Uniquat (UPL Ltd.), Hanuxone (Dhanuka Agritech Ltd.), Weedax (Coromandel), Swat (Shaw Wallace), Paralac (Rallis)

Herbicide	***Brand Names and Manufacturers***
Pendimethalin	38.7% CS: Stomp xtra (BASF), Pendigold (Syngenta), Tata Ponida (Rallis), Bunker (PI Ind.), Stop (Krisi Rasayan), Gaadar (Hindustan Pulverising), Anustorm (Anu), Pendiguard (Gharda), Penadril (Crystal), Pendiherb (Cheminova), Dhanustomp (Dhanuka Agritech Ltd.), Penda (Bharat Insecticide), Excel Plus (Excel), Speed (Indofil), Nagastra (Nagarjuna), Eezykill (Coromandel), Swalpendi (Shaw Wallace)
	5% Gr : Prowl (BASF), Klean up (Punjab Chemicals)
Pretilachlor	50% EC: Rifit (Syngenta), Pretiherb (Cheminova), Remove (Biostadt), Delete (Godrej), Erase (Nagarjuna), Offset (Indofil), Tatapreet (Rallis), Hifit (Hindustan Pulverising), Blade (Gharda), Sureshot (Sulfur Mills), Shri Ram pretilachlor (Shri Ram Chem), Alchor (PI Ind.), Pretit (FIL Ind.), Pilot (EID Parry), Pilot (Coromandel), Excel merit (Excel), Errant (Bharat Rasayan), Razor (Plant Remedies), HIL-Preti (HIL), Fiesta (FMC), Prince (Krishi Rasayan), Sift and Growfit (Crystal), Weep (Hem-an), Pretobest (DuPont),
	Crase (Dhanuka Agritech Ltd.), Giant (ECI), Nag Pretichlor (Multiplex), Racer (Insectcides India), Profit (Devidayal), Phenom (Sudershan), Sicle (GSP), Lorette (Coromandel), Zinta (Heranba), Dismiss (Isagro), Pretigan (Makhsteshim-Agan), TKS Pretila (Tata Chem), Klean Out (Punjab Chemicals), Pitla (Shaw Wallace), Preet (Rallis), Vinia (Atul)
	30.7%EC: Erase-N (Nagarjuna), Errant-N (Bharat Rasayan), Sofit (Syngenta), Pretigan-S (ADAMA), Wranglar (Super Agro)
	30% EW: Erasan (Syngenta)
Pinoxaden	5% EC: Axial (Syngenta)
Propanil	35% EC: Stam F-34 (Indofil), Dhanustam (Dhanuka Agritech Ltd.), Melsa (PI Ind.)
Propaquizafop	10% EC: Society (Indofil)
Pyrazosulfuron-ethyl	10% WP: Saathi (UPL Ltd.)
Pyrithiobac-sodium	10% EC: Theme (DuPont), Rife (Cheminova)
Quizalofop-ethyl	5% EC: Targa Super (Dhanuka Agritech Ltd.), Assure (DuPont)
Quizalofop-P-tefuryl	4% EC: Pantara (DuPont)

Herbicide	***Brand Names and Manufacturers***
Sulfosulfuron	75% WP: Sulfo and SF 10 (UPL Ltd.), Safal (Gharda), Décor (Indof il), Don (Herbicide India), Leader (Sumitomo), Nagsuron (Multiplex),Target (Hindustan Pulverising), Fateh (Rallis), Wazir (Bharat Rasayan), Arrow (Sulfur Mills), Top shot (Hem-an), Kaizer and Guru (Insecticide India), Safari (Devidayal), Clear (GSP), Razor or Kripon (Crystal), Sumet (Cheminova), Excel-Lolkar (Excel), Toss (JU Pesticid e), Sulfom en (ADAMA), sultop (Dhanuka Agritech Ltd.), Image (Heranba), Blanket (Punjab Chemicals), Unik (Shaw Wallace), Loxo (Atul)
Tembotrione	34.4% SC: Laudis (Bayer)
Thiobencarb (Benthiocarb)	50% EC: Sattern (PI Ind.), Dhanustern (Dhanuka Agritech Ltd.)
	10% Gr: Thiobencarb (Vimal)
Topramezone	33.6% SC: Tynzer (BASF)
Triallate	50% EC: Avadex-BW (Monsanto)
Trifluralin	48% EC: Flora (UPL Ltd.), Clean (Bayer), Tiptop (Gharda)
Triasulfuron	20% WG: Logran (Syngenta)
Chlorimuron-ethyl + Metsulfuron-methyl	20% WP: Almix (DuPont), Pimix (PI Ind.), Cormix (Coromandel)
Anilophos + 2,4-D ethyl ester	56% EC: Arozine D (Bayer), Topshort (Gharda)
Bensulfuron + Pretilachlor	6.6% G: Londex Power (DuPont), Rizal (Super Agro)
Clodinafop + Metsulfuron- methyl	16% WP: Sandesh (Shaw Wallace), Vesta (UPL Ltd.)
Hexazinone + Diuron	60% WP: Velpar K-4 (DuPont)
Sulfosulfuron + Metsulfuron-methyl	80% WG: Total (UPL Ltd.), SataSat (Shaw Wallace)
Imazamox + Imazethapyr	70% WG: Bingo (PI Ind.), Pyramox (Coromandel)
Carfentrazone-ethyl + Sulfosulfuron	45% WG: (FMC India Pvt. Ltd.)
Sodium acefluorfen + Clodinafop-propargyl	24.5% EC: Irish (UPL Ltd.)
Imazethapyr + Pendimethalin	32% EC: Valor (BASF)

Source: Directorate of Plant Protection, Quarantine & Storage, Central Insecticides Board & Registration Committee, Faridabad, 2020

Chapter 7

Important Agronomy Formulas

Weed Management

Relative Density (RD)

The relative density (RD) of weeds was worked out using the following formula

$$RD(\text{per cent}) = \frac{\text{No. of weeds of individual species}}{\text{Total Number of weeds}} \times 100$$

Relative Dry Weight (RDW)

$$RDW\ (\text{per cent}) = \frac{\text{Dry weight of given species}}{\text{Total dry weight of all species}} \times 100$$

Weed Control Efficiency (WCE)

Weed control efficiency was computed by adopting a formula suggested by Mani *et al.* (1973) and expressed in percentage.

$$WCE = \frac{WD_C - WD_T}{WD_C} \times 100$$

where,

WD_C = Weed density (number m^{-2}) in the control plot

WD_T = Weed density (number m^{-2}) in the treated plot

Weed Index (WI)

Weed index was calculated by following the formula suggested by Gill and Vijayakumar (1966).

$$WI=\frac{X-Y}{X}\times 100$$

where,

X = Yield from plot with minimum weeds (kg ha^{-1})

Y = Yield in the respective treatment plot (kg ha^{-1}) for which the weed index to be calculated

Summed Dominance Ratio (SDR)

The major or dominant weed species were determined by computing SDR values as given by Bhandari (1981),

Absolute density (D) = The total number of plants for a given species in all quadrates

$$\text{Relative density (Rd)} = \frac{\text{Absolute density for a species (D)}}{\text{Total number of plants for all species}}\times 100$$

$$\text{Absolute frequency(F)} = \frac{\text{The number of quadrates in which a given species occurs}}{\text{Total number of quadrates used (10)}}\times 100$$

$$\text{Relative frequency(Rf)} = \frac{\text{Absolute frequency for a species (F)}}{\text{Total of the absolute frequencies for all species}}\times 100$$

Importance Value (I.V.) Index = Relative density (Rd) + Relative frequency (Rf)

$$\text{Summed dominance ratio (SDR)} = \frac{\text{I.V.}}{2}$$

Calibration of a Sprayer

The process to find out the exact quantity of carrier (water) required for spraying a given area under a set of conditions is called "calibration". Calibration of a sprayer is a must for spraying herbicide in the field. Care must be taken to see that speed and pressure do not vary significantly from the calibration test to actual field spraying. Calibration can be done by filling a known quantity of water into the sprayer and spraying it over a measured area.

The quantity of the water required for spraying a given area can be calculated by using the following formula.

$$Q = \frac{(V_1 - V_2)}{a} \times A$$

where,

Q = Quantity of water required for area to be sprayed (liters)

V_1 = Quantity of water taken initially

V_2 = Quantity of water remained after spraying

A = Area to be sprayed m^2

a – Measured area sprayed m^2

The area covered per hour can be calculated as shown under

$$\text{Area (ha)/hr} = \frac{\text{Walking speed (km/hr)} \times \text{m/km} \times \text{Spray width (m)}}{\text{m}^2\text{/ha}}$$

Calculating Commercial Formulation

The commercial formulation of herbicide required can be worked out using the formula

$$\text{Formulated product required} = \frac{\text{Rate of a.i/ ha to be applied} \times 100}{\text{Active ingredient (per cent)}}$$

For example, if one needs to apply Diuron @ 2 kg a.i./ha and the commercial formulation contains 80 per cent a.i., the total product required per ha would be:

$$= \frac{2 \times 100}{80} = 2.5 \text{ kg}$$

Irrigation Management

Water Productivity

Water productivity is a function of total water consumed and gross return obtained by the crop and expressed in m^{-3}.

$$\text{Water productivity} = \frac{\text{Gross return}}{\text{Total water applied (m}^{-3}\text{)}}$$

Total Water Consumed

The total water consumed was computed by summing the irrigation water applied and the effective rainfall. Effective rainfall calculated as fifty percentage of total rainfall during the cropping period.

W = ND + Re

where,

W = Total water consumed in mm

N = Number of irrigations

D = Applied water depth for each irrigation (mm)

Re = Effective rainfall (mm), during the cropping period

Irrigation Efficiency

It is defined as the ratio of water output to the water input, *i.e.* the ratio or percentage of the irrigation water consumed by the crop of an irrigated farm or field or project to the water delivered from the source.

$$E_i = \frac{W_c}{W_f} \times 100$$

where,

E_i = Irrigation efficiency (per cent)

W_c = Irrigation water consumed by crop during growth period in an irrigation project.

W_r = Water delivered from project during the growth period of crops.

In most of the irrigation projects, the irrigation efficiency ranges between 12 to 34 per cent.

Water Conveyance Efficiency

It is a measure of efficiency of water conveyance system from canal network to watercourses and field channels. It is the ratio of water delivered infields at the outlet head to that diverted into the canal system from the river or reservoir. Water losses occur in conveyance from the point of diversion till it reaches the farmer's fields can be evaluated by water conveyance efficiency, as under:

$$E_c = \frac{W_f}{W_d} \times 100$$

where,

E_c = Water conveyance efficiency (per cent)

W_f = Water delivered to the farm by conveyance system (at field supply channel)

W_d = Water introduced into the conveyance system from the point of diversion

Water conveyance efficiency is generally low; about 21 per cent losses occur in earthen water courses only.

Water Application Efficiency

It is a measure of efficiency of water application in the field. It is the ratio of volume of water that is stored in the root zone of crops and ultimately consumed by transpiration or evaporation or both to the volume of water actually delivered at the field.

Alternatively, it may be defined as the percentage of water applied that can be accounted for an increase in soil moisture in soils as occupied by the principal rooting system of the crop. It is also termed as farm efficiency as it takes into account water lost in application at the farm.

$$E_a = \frac{W_s}{W_f} \times 100$$

where,

E_a = Water application efficiency (per cent)
W_s = Irrigation water stored in the root zone of farm soil
W_f = Irrigation water delivered to the farm (at field supply channel)

In general, water application efficiency decreases as the amount of water during each irrigation increases. Water losses due to inefficient application of water in the field vary from 28 to 50 per cent.

Water Use Efficiency

$$E_u = \frac{W_u}{W_d} \times 100$$

where,

E_u = Water use efficiency (per cent)
W_u = Water beneficially used
W_d = Water delivered

It is expressed in kg/ha cm

Water use efficiency is also defined as (i) crop water use efficiency and (ii) field water efficiency.

(a) Crop Water Use Efficiency

It is the ratio of yield of crop (Y) to the amount of water depleted by crop in evapotranspiration (ET).

$$CWUE = \frac{Y}{ET}$$

where,

CWUE = Crop water use efficiency
Y = Crop yield
ET = Evapotranspiration

It is expressed in kg/ha/mm or kg/ha/cm.

(b) *Field Water Use Efficiency*

It is the ratio of yield of crop (Y) to the total amount of water used in the field.

$$FWUE = \frac{Y}{WR}$$

where,

FWUE = Field water use efficiency

WR = Water requirement

It is expressed in kg/ha/mm (or) kg/ha/cm

Water Storage Efficiency

It is defined as the ratio of the water stored in the root depth by irrigation to the water needed in the root depth to bring it to the field capacity. Also termed as water storage factor.

$$E_s = \frac{W_s}{W_w} \times 100$$

where,

E_s = Water storage efficiency (per cent)

W_s = Water stored in the root zone during the irrigation

W_w = Water needed in the root zone prior to irrigation, *i.e.*, field capacity available moisture.

Water Distribution Efficiency

Expression for distribution efficiency to evaluate the extent to which the water is uniformly distributed is as follows:

$$E_d = \left(1 - \frac{d}{D}\right)100$$

where,

E_d = Water distribution efficiency (per cent)

d = Average numerical deviation in depth of water stored from average depth stored during irrigation

D = Average depth of water stored along the run during irrigation

A water distribution efficiency of 80 per cent means that 10 per cent of water was applied in excess and consequently 10 per cent was deficient in comparison to the average depth of application.

Green and Blue Water

Green water use refers to the volume of rainwater that evaporates from a crop field during the growing period. Blue water use refers to the volume of irrigation water (withdrawn from surface or ground water) that evaporates from a crop field during the growing period. The green and blue water was calculated as per the procedure presented by (Falkenmark, and Rockström., 2004). The green and blue water crop water use is calculated as follows.

$$CWU_{green} = 10 \times \sum_{D=1}^{1gp} ET_{green}$$

$$CWU_{blue} = 10 \times \sum_{D=1}^{1gp} ET_{blue}$$

The green and blue components in crop water use (CWU, m^{-3} ha^{-1}) was calculated by accumulation of daily evaporation (ET, mm/day) over the complete growing period in which ET green represents green water evapotranspiration and ET blue represents blue water evapotranspiration. The factor 10 is mean to convert water depths in mm into volume per land surface in m^{-3} ha^{-1}. The summation is done over the period from the day of planting (day 1) to the day of harvest (lgp stands for length of growing period in days).

Drip Irrigation

System Capacity Calculation

Discharge rate of the pump = 6 liters/sec

Normal pumping duration = 3 hours/day

Peak evaporation rate = 5 mm/day

The Area that can be Irrigated by the System

$$\text{Area} = \frac{\text{Flow rate (lph)} \times \text{Pumping duration (hrs/day)}}{10000 \times \text{Gross irrigation requirement (mm/day)}}$$

Discharge rate = 6 liters/minute

6 x 60 = 360 liters

6 x 360 = 21600 liters/hours

where,

$$\text{Total Area to be irrigated} = \frac{21600 \times 3}{10000 \times 5} = 1.30 \text{ ha/day or } 3.25 \text{ acre/day}$$

Water Requirement

a) *Number of Emitter/ha in to be Calculated*

$$AE = \frac{10000}{\text{Lateral spacing (m)} \times \text{Emitter spacing (m)}}$$

Normal lateral spacing = 120 cm or 1.2 m

Normal emitter spacing = 45 cm or 0.45 m

$$\text{Number of emitters} = \frac{10000}{1.2\,\text{m} \times 0.45\,\text{m}} = 18518\ \text{emitter/ha}$$

Calculate the Discharge Rate

Discharge rates of drippers in 4 lph

Water discharged rate = 74080 liters/ha or 29632 liters/acre

Nutrient Management

Nitrogen Use Efficiency

Various parameters used to study the use efficiency of nitrogen are as follows.

Agronomic Efficiency

The agronomic efficiency *i.e.* the response in yield per unit input as indicated by (Yoshida, 1981).

$$AE = \frac{\text{Grain yield in fertilised plot (kg ha}^{-1}\text{)} - \text{Grain yield in unfertilized plot (kg ha}^{-1}\text{)}}{\text{Quantity of fertilizer N applied (kg ha}^{-1}\text{)}}$$

Physiological Efficiency

The physiological efficiency also known as efficiency of utilization as indicated by kg of grain per kg of absorbed N (Yoshida, 1981) was computed as follows

$$PE = \frac{\text{Grain yield in fertilised plot (kg ha}^{-1}\text{)} - \text{Grain yield in unfertilized plot (kg ha}^{-1}\text{)}}{\text{N uptake in fertilized plot (kg ha}^{-1}\text{)} - \text{N uptake in unfertilized plot (kg ha}^{-1}\text{)}}$$

Apparent N Recovery

Apparent N recovery, also known as recovery fraction was computed as per the formula suggested by Pillai and Vamadevan (1978).

$$ANR = \frac{Y_t - Y_o}{N_t} \times 100$$

where,

Y_t = Uptake of N in particular treatment (kg ha^{-1}),

Y_0 = Uptake of N in unfertilized plot (kg ha^{-1}), and

N_t = Quantity of N applied for the treatment (kg ha^{-1}).

Energy Analysis

The required data from the experiment were collected and computed using the method suggested by Kalbande and More (2008) and Devasenapathy *et al.* (2006)

Energy Efficiency (EE)

Cultural energy utilized through inputs and energy produced as products are calculated and expressed in Mega Joules (MJ). Energy efficiency was worked out taking in account the input and output energy for each treatment (Dazhong and Pimental, 1984).

$$EE = \frac{\text{Energy output (MJ/ha)}}{\text{Energy input (MJ/ha)}}$$

Specific Energy (SE)

Specific energy of the treatment was calculated in terms of energy required to produce one kg of main product and expressed in MJ/kg

$$SE = \frac{\text{Energy input (MJ/ha)}}{\text{Grain/economic yield (kg/ha)}}$$

Net Energy (NE)

The net energy of each crop was calculated by deducting the energy input from the energy output of particular treatment or practice.

NE = Energy output (MJ/ha) – Energy input (MJ/ha)

Energy Productivity (EP)

Energy productivity is the quantity of physical output obtained for every unit of input and expressed as kg/MJ.

$$EP = \frac{\text{Output (grain + byproduct) (kg/ha)}}{\text{Energy input (MJ/ha)}}$$

Energy Intensity (EI)

Energy Intensity is the ratio between energy output and cultivation expenses and expressed as MJ/Rs.

$$EI = \frac{\text{Energy output (MJ/ha)}}{\text{Cost of cultivation (Rs/ha)}}$$

Indices Based on Land Use Systems

1. Multiple Cropping Index (MCI): Dalrymple (1971)

It measures the sum of areas planted and harvested to different crops in a single year divided by total cultivated area times 100.

$$MCI = \frac{\sum_{i=1}^{n} ai}{A} \times 100$$

where,

n = Total no. of crops

ai = Area occupied by i^{th} crop

A = Total land area

2. Cropping Intensity/Intensity of Cropping (CI)

$$CI = \frac{\text{Total cropped area (Gross cropped area)}}{\text{Net cultivated area}} \times 100$$

$$CI = \frac{\text{Area under kharif + rabi + summer crops}}{\text{Area under actual cultivation}} \times 100$$

3. Rotational Intensity (RI)

$$RI = \frac{\text{No. of crops grown in a field}}{\text{Years of rotation}} \times 100$$

e.g., 1. Maize– potato – onion: 1 year

$$RI = \frac{3}{1} \times 100 = 300 \text{ per cent}$$

2 .Maize – sugarcane – onion: 2 years

$$RI = \frac{3}{2} \times 100 = 150$$

4. Cropping Intensity Index (CII): (Menegay *et al.*, 1978)

It assesses a farmer's actual land use in area and time relationships for each crop or group of crops compared to the total available land area and time, including land temporarily available for production. Efficient cropping zone is judged by CII and LER

$$CII = \frac{\sum_{i=1}^{N_C} a_i\, t_i}{\sum_{j=1}^{M} A_i\, T_i}$$

where,

N_c = Total no. of crops grown during time period T

A_i = area occupied by i^{th} crop

ti = duration occupied by i^{th} crop (months that crop occupied ai)

T = Time period (usually one year)

A_o = Total cultivated land area available for use during T

M = Total no. of fields temporarily available during T

A_j = Land area of jth field

T_j = Time period when Aj is available

5. Specific Crop Intensity Index (SCII): (Menegay *et al.,* 1978)

It is the derivative of CII and determines the amount of area-time denoted to each crop or group of crops compared to the total time available

$$\text{SCII} = \frac{\sum_{k=1}^{Nk} a_k\, t_j}{\sum_{j=1}^{M} A_j\, T_j}$$

N_k = Total no. of crops during T

a_k = Area occupied by K^{th} crop

t_k = Duration of Kth crop

6. Relative Cropping Intensity Index (RCII)

It is the modification of CII and determines the amount of area-time allotted to one crop or group of crops relative to the area-time actually used in the production of all the crops in a farm.

$$\text{RCII} = \frac{\sum_{k=1}^{Nk} a_k\, t_j}{\sum_{j=1}^{M} a_{ki}\, t_i}$$

It is used for classifying farmers *viz.,* when relative vegetable Intensity Index is more than 50 per cent then farmers will be called vegetable grower.

7. Cultivated Land Utilization Index (CLUI): Chuang (1973)

It is calculated by summing the products of land area planted to each crop, multiplied by the actual duration of that crop and divided by the total cultivated land area times 365 days.

$$\text{CLUI} = \frac{\sum_{i=1}^{n} a_i\, d_i}{A \times 365}$$

where,

i = 1, 2, 3 crops 1....n.

n = Total no. of crops

a = Area occupied by i^{th} crop.

d = Days that i^{th} crop occupied ai.

A = Total cultivated land area available during 365 days period.

8. Area Time Equivalent Ratio (ATER)

It takes into account the duration of crops and permits an evaluation of crops on yield per day basis. It is a modification of LER.

$$\text{ATER} = \frac{(\text{LA} \times \text{DA}) + (\text{LB} \times \text{DB})}{\text{T}}$$

where, LA and LB are relative yield or partial LER of component crops. A and B, DA and DB are duration of crops. A and B, and T are the total duration of the intercropping system.

9. Staple Land Equivalent Ratio (SLER)

In situations, where the primary objective is to produce fixed yield of one component (staple) and some yield of other crop, the concept of SLER is proposed.

$$\text{SLER} = \frac{\text{MDA}}{\text{SA}} + \text{P}\left(\frac{\text{MB}}{\text{SB}}\right)$$

where, MDA is derived yield of A in mixture yield and P the proportion of land devoted for intercropping. The two crop yields in the intercrop mixture are MA and MB and the yield of crops grown as sole crop are SA and SB.

Indices Based on Biological Potential

Crop Equivalent Yield

The equivalent yield should preferably be calculated in terms of kharif crop (say paddy in rice-based cropping system) using the following formula.

$$\text{CEY} = \frac{\text{Yi} \times \text{Pi}}{\text{Pb}}$$

where,

CEY = Crop equivalent yield

Yi = Yield of individual crop

Pi = Price of individual crop

Pb = Price of base crops

Example: Yield and cotton equivalent yield of crops in different sequential cropping systems

Crop Sequence	*Grain Yield (kg/ha)*				*Cotton Equivalent Yield (kg/ha)*			
	K	*R*	*S*	*Total*	*K*	*R*	*S*	*Total*
Cotton -Finger millet- Sorghum	1903	2833	2460	7196	1903	708	738	**3349**
Beetroot - Green gram - Maize	9937	1035	5626	**16598**	2484	1035	1969	**5488**
Cowpea - Gingelly - Maize	962	762	5246	**6970**	481	1524	1836	**3841**
Cotton – *S.culeate* - Sunflower	1656	12054	1633	**15343**	1656	301	1388	3345

Produce Market Prices (Rs./kg)

Cotton: Rs. 20 Finger millet: Rs. 5 Sorghum: Rs. 6 Beetroot: Rs. 5

Green gram:Rs. 20 Maize: Rs. 7 Cowpea: Rs. 10 Gingelly: Rs.40

S.aculeate: Rs. 0.50 Sunflower: Rs. 17

One other form of single measurement comparison which is exactly equivalent to the financial value index, is the crop equivalent.

In calculating a crop equivalent, yield of one crop is converted into yield equivalent of the other crop by using the ratio of prices of the two crops.

Indices based on Biological potential are categorized into production efficiency index and Interference indices.

I. Production Efficiency Index

1. LER (Land Equivalent Ratio)

LER is the relative land areas under sole crop required to produce the same yield as obtained under a mixed or inter cropping system at the same level of management.

It is the ratio of land required by pure crop to produce the same yield as that of intercrop.

$$LER = \frac{Ya}{Sa} + \frac{Yb}{Sb}$$

where,

Ya and Yb = Yield of individual crops 'a' and 'b' respectively in mixture.

Sa and Sb = Yields of individual crops 'a' and 'b' respectively in pure stand.

Thus, it gives a better picture of the competitive abilities of the component crops. It also gives actual yield advantage of intercropping.

In another words LER is the unit to measure the production efficiency of different intercropping system by converting the production in terms of land acreage. LER gives an accurate assessment of the biological efficiency of intercropping.

When LER > 1, intercropping is beneficial.

2. Relative Yield Total and Land Equivalent Ratio

The most important index of biological advantage is the relative yield total (RYT) introduced by De Wit and Van Den Bergh (1965) or land equivalent ratio (LER) reviewed by Willey (1979).

The mixture yield of a component crop expressed as a portion of its yield as a sole crop from the same replacement series is the relative yield of the crop and the sum of the relative yields of component crops is called relative yield total (RYT). The total land area required under sole cropping to give the same yields obtained in the intercropping mixture is called land equivalent ratio (LER). Both the expressions (RYT and LER) are similar.

Example

Yield of groundnut and pigeon pea as sole crops were 1000 and 500 kg ha^{-1}. Corresponding yield in intercropping was 700 and 400 kg $ha^{-1.}$ Price of pigeon pea is Rs.12 kg –I and that of groundnut Rs.12 kg^{-1}. Calculate LER.

Solution

$$\text{LER} = \frac{700}{1000} + \frac{400}{500} = 1.5$$

There are two different objectives for which such indices have been proposed.

The first is the assessment of the benefit or overall advantage of intercropping and the second is the assessment of the relative performance of the two crops, the concept of dominance or competitiveness.

It is important not to confuse these two objectives, which should be quite separate conceptually.

The RYT or LER is the main index of advantage currently used.

II. Interference Efficiency

1. Relative Crowding Coefficient (RCC) Proposed by De Wit (1960)

It is used in replacement series of intercropping.

It indicates whether a species or a crop, when grown in mixed population, has produced more or less yield than expected in pure stand.

In 50:50 mixture

$$\text{Kab (RCC)} = \frac{\text{Mixture yield of a}}{\text{Pure stand yield of a} - \text{Mixture yield of a}} = \frac{Yab}{Yaa - Yab}$$

where,

Yab = Mixture yield of crop 'a' grown with 'b'

Yba = Mixture yield of crop 'b' grown with 'a'

Yaa = Yield in pure stand of crop 'a'

Ybb = Yield in pure stand of crop 'b'

Zab = Proportion of sown species 'a' in mixture with 'b'

Zba = Proportion of sown species 'b' in mixture with 'a'

For all mixture

$$\text{Kab} = \frac{Yab \times Zba}{(Yaa - Yab)Za}$$

K > 1 Means yield advantage (more yield than expected)

K = 1 No difference

K < 1 Yield disadvantage (less yield than expected)

Crowding coefficient and LER give the yield advantage but only LER given the magnitude of advantage. Therefore, LER is preferred to assess the competition effects and yield advantage in intercropping situations.

2. Aggressivity

Proposed by Mc Gihrist (1965). It is the mixture of how much the relative yield increase in component 'a' is greater than that for component 'b'.

$$\text{Kab} = \frac{\text{Mixture yield of a Mixture yield of b}}{\text{Expected yield of a Expected yield of b}}$$

$$\text{Aab} = \frac{Yab}{Yaa \times Zab} - \frac{Yba}{Ybb \times Zab}$$

where,

Aab = 0 means component crops are equally competitive

Aab = negative means dominated

Aab = bigger value either positive or negative means bigger difference in competitive abilities.

3. Competition Index (CI): Donald (1963)

It is a measure to find out the yield of various crops when grown together as well as separately. It indicates the yield per plant of different crops in mixture and their respective pure stand on a unit area basis.

If the yield of any crop, grown together is less than its respective yield in pure stand then it is harmful association but on increased yield means positive benefit.

$$\text{CI} = \frac{(Yaa - Yab)\times(Ybb - Yba)}{Yaa \times Yab}$$

4. Competition Ratio (CR)

Proposed by Willey and Rao (1980)

$$\text{Cra} = \frac{Yab}{Yaa \times Zab} + \frac{Yba}{Ybb \times Zab}$$

It is simply the ratio of individual LERs of the two component crops, but correcting for the proportion in which they were initially sown.

5. Competition Coefficient (CC)

Ratio of the Relative crowding coefficient (RCC) of any given species in the mixture.

$$\text{CC} = \frac{\text{RCC of given spp}}{\text{Total RCC of all crop in mixture}}$$

It is used to find out the relative crowding from which maximum yield can be obtained without any adverse effect on any of the species.

Indices Based on Economic Viability

1. Gross Return

The total monetary returns of the economic produce such as grain, tuber, bulb, fruit, *etc.* and by products *viz.*, straw, fodder, fuel, *etc.* obtained from the crops included in the system are calculated based on the local market prices. The total return is expresses in terms of unit area, usually one hectare.

2. Net Return or Net Profit

Worked out by subtracting the total cost of cultivation from the gross returns. This value gives the actual profit obtained by the farmer. In this type of calculation only the variable costs are considered

3 Returns/Rupee Invested

Also called as benefit cost ratio or input output ratio.

$$\text{Return per rupee invested} = \frac{\text{Gross returns (Rs.)}}{\text{Cost of cultivation (Rs.)}}$$

4. Per Day Return

Also called as income per day and can be obtained by dividing the net

$$\text{Per day return} = \frac{\text{Net returns (Rs.)}}{\text{Cropping period (days)}}$$

5. Diversity Index (DI): Strout and Wang and Yu (1975)

It measures the multiplicity of crops or farm products which are planted in a year by computing the reciprocal of sum of squares of the share of gross revenue received from each individual farm enterprise in a single year.

In the following farms, which farm is mostly specialized.

Crops	*Income (Rs.) from Different Farms*		
	A	*B*	*C*
Sugarcane	30,000	-	10,000
Cotton	10,000	20,000	20,000
Wheat	40,000	29,000	10,000
Jowar	20,000	10,000	40,000
Potato	-	50,000	-
Total	**100,000**	**100,000**	**80,000**

Solution

Crops	*Shares of individual crops in different Farmer Participatory research on Integrated Farming System in Low External Input Sustainable Environment*					
	Farm A		*Farm B*		*Farm C*	
	Share	*Square of its Share*	*Share*	*Square of its Share*	*Share*	*Square of its Share*
Sugarcane	0.3	0.09	-	-	0.125	0.0156
Cotton	0.1	0.01	0.2	0.04	0.25	0.625
Wheat	0.4	0.16	0.2	0.04	0.125	0.0156
Jowar	0.2	0.04	0.1	0.01	0.50	0.2500
Potato	-	-	0.5	0.25	-	-
Total	**1.0**	**0.30**	**1.00**	**0.34**	**1.00**	**0.3437**

Diversity Index

$$\text{Farm A} = \frac{1.00}{0.30} = 3.33$$

$$\text{Farm B} = \frac{1.00}{0.34} = 2.94$$

$$\text{Farm} = \frac{1.00}{0.3437} = 2.91$$

Therefore, Farm 'A', is more diversified and Farm C is more specialized Lower is the diversity Index, higher is the specialization.

$$\text{HDI} = \frac{1}{\sum_{i=1}^{n} yi \sum_{i=1}^{n} yi}$$

where,

yi = Gross value of the ith crop planted and harvested within a year.

Biometric Observations

Germination Percentage

$$\text{Germination Percentage} (\%) == \frac{\text{Number of germinated seeds}}{\text{Total number of seeds sown}} \text{ x } 100$$

Germination Index

$$\text{Germination index} = \frac{\text{No.of germinating seeds}}{\text{Days of first count}} + \frac{\text{No.of germinating seeds}}{\text{Days of final or last count}}$$

Vigour Ondex

Vigour index = Germination percent x seedling total length

Rate of Emergence

No. of seedlings emerged 5 days after sowing

Rate of emergence = -- x 100

No. of seedlings emerged 15 days after sowing

Relative Leaf Water Content (RLWC)

$$\text{RWC} (\%) = \frac{\text{Fresh Weight} - \text{Dry Weight}}{\text{Turgid Weight} - \text{Dry Weight}} \text{ x } 100$$

Leaf Area Index (LAI)

$$LAI = \frac{L \times W \times K \times \text{Number of leaves plant}^{-1}}{\text{Land area occupied by plant}}$$

where,

L = Length of leaf (cm)

W = Width of leaf (cm)

K = Correction factor (based on crops and season).

Leaf Area Duration (LAD)

$$LAD = \frac{L_1 + L_2}{2} \times (t_2 - t_1)$$

where,

L_1 and L_2 are the LAI at time t_1 and t_2.

Crop Growth Rate (CGR)

The CGR was expressed in g m^{-2} day^{-1}.

$$CGR = \frac{W_2 - W_1}{P(t_2 - t_1)}$$

where,

W_1 and W_2 are whole plant dry weight (g) at time t_1 and t_2, respectively

t_1 and t_2 are the initial and final day of period of observation, respectively

P is the plant spacing adopted (m^2)

Relative Growth Rate (RGR)

The RGR was expressed in mg g^{-1} day^{-1}.

$$RGR = \frac{\text{Log}_e W_2 - \text{Log}_e W_1}{(t_2 - t_1)}$$

where,

W_1 and W_2 are the initial and final dry weight of plants, respectively

t_1 and t_2 are the initial and final day of period of observation, respectively

Specific Leaf Weight (SLW)

The SLW was expressed in g cm^{-2} at panicle initiation stage.

$$SLW = \frac{\text{Leaf dry weight (g)}}{\text{Total leaf area (cm}^2\text{)}}$$

Number of Productive Tillers m^{-2}

$$\text{Number of productive tillers m}^{-2} = \text{Number of productive tillers hill}^{-1} \times \text{plant population m}^{-2}$$

Fertility Percentage

$$\text{Fertility percentage} = \frac{\text{Number of filled grains per panicle}}{\text{Total number of grains per panicle}} \times 100$$

Chapter 8

Agronomy Point to Points

- ☆ Brown manuring a resource conservation technology is commonly employed in: **Rice**
- ☆ Botanical name of Ragi is: ***Eleusine coracana***
- ☆ Botanical name of Sewan grass is: ***Lasiurus sindicus***
- ☆ Total annual loss in agricultural produce due to weeds accounts for: **45 per cent**
- ☆ The largest producer of moth bean in India is: **Rajasthan**
- ☆ A cloud type that occurs in low height in the atmosphere is: **Stratus**
- ☆ Crop suitably grown as a mixed crop with wheat is: **Mustard**
- ☆ Costing is when you done in field: **Keep making plough turn to the right**
- ☆ Most critical stage of irrigation in potato is: **Stolon Formation**
- ☆ Ramie (*Boehmeria nivea*) is a crop: **Fibre crop**
- ☆ Late planting of autumn sugarcane reduces the germination due to: **Low temperature**
- ☆ Proper stage of incorporation of Green manuring crop sunnhemp is: **4-5 weeks after sowing**
- ☆ Which mineral is called Fool's Gold: **Iron Pyrite**
- ☆ When air temperature and dew point are equal: **100 per cent**

- ☆ Dehaulming in potato is done to obtain: **Improve the quality of seed tuber**
- ☆ Practice of cross ploughing the young crop of rice to reduce weed and vegetative growth of the crop is called: **Beushening**
- ☆ Critical temperature below which pollen sterility is induced in paddy: **12^0C**
- ☆ Effective method of biofertilizers application in the transplanted crops is: **Root dip method**
- ☆ Multistoried cropping system is practiced in: **Karnataka and Kerala**
- ☆ The Seed rate of Sesame is: **3-4 kg**
- ☆ A value concept was given by: **Fried and Dean**
- ☆ Which one of the following is best example of catch crops: **Toria**
- ☆ The Fat content of bajra is: **5 per cent**
- ☆ CO_2 content in soil air is: **0.25 per cent**
- ☆ Berseem was introduced in India from in 1904: **Egypt**
- ☆ Inter-culture in groundnut is avoided at: **Pegging Stage**
- ☆ Seed work board is required for: **Blending Test**
- ☆ The White Revolution in India was brought by: **V. Kurien**
- ☆ The oil and protein content of groundnut are: **45 per cent and 26 per cent**
- ☆ Which one is not a kharif season weed: ***Chenepodium album***
- ☆ Which one is an apophytes weed (introduced by man from other place): ***Phalaris minor***
- ☆ Which one is an example of 'Satellite weed' (travelled along with harvested crops): ***Avena fatua***
- ☆ A device for measuring percolation and leaching losses from a column of soil under controlled conditions is known as: **Lysimeter**
- ☆ At Permanent Wilting Point (PWP) the pF value is: **4.2**
- ☆ Tetrazolium test is used to determine: **Seed viablity**
- ☆ The instrument used for measuring depth of water table is known as: **Piezometer**
- ☆ UPAS 120 is a variety of: **Pigeon pea**
- ☆ The optimum seed rate for normal sown crop of lentil is (kg/ha): **30-40 kg**

- Black gram is originated in: **India**
- Bund former is used for making: **Soil smooth**
- Criteria of Essentiality of nutrient was given by: **Arnon and Stout**
- 'Tarai' soil are deficient in: **Zn**
- Soil sickness is caused by: **Jowar and Linseed**
- Weight of one cotton bale is equal to: **170 kg**
- The Ideal stage of harvest of jute for fiber purpose is: **Small Pod Stage**
- Highest procurement of wheat in 2009 was Million tons: **17.8**
- Farming system is: **All agricultural Inputs and commodities**
- Which nutrient helps in Biological Nitrogen Fixation: **Mo**
- Silt is intermediate between: **Sand and clay**
- CIMMYT works on: **Maize and wheat**
- Price fixed by government recently for sugarcane is: **Fair and Remunerative Price**
- If farmer has only one irrigation is available for wheat crop, at which stage it is recommended: **Crown Root Iniation (CRI)**
- Widely cultivated wheat species in India after *T. durum*: ***T. diococcum***
- Total geographical area of India: **328 Mha**
- Family of cotton is: ***Malvaceae***
- Which of the following is used to turn over the soil: **Mould bould plough**
- Puddled soil has higher water use efficiency than puddle soil: **2.5 per cent**
- Secondary tillage done only in seed rows is: **Row zone tillage**
- Commonly used herbicide under zero tillage is: **Diquat and Paraquat**
- In which part of India rice is grown throughout the year: **South and East**
- SUBACAS is term related to: **Sugarcane**
- Most prominent cropping system of India is: **Rice- Wheat**
- Rice seed dormancy can be broken by: **Heat treatment and Nitric acid treatment**
- Ultimate potential of irrigation in India is: **140 mha**
- River having highest ground water potential: **Ganga**

- ☆ Rapid means of *in-situ* measurement of soil moisture is made through: **Tensiometer**
- ☆ Relationship between crop yield and ET in pulse is: **Quadratic**
- ☆ Percentage of water used for metabolic process of crop is: **<1 per cent**
- ☆ 1 ha cm of irrigation is equal to: **100000 liters**
- ☆ Flooding method of irrigation is suitable for Crop: **Rice**
- ☆ Leading state in use of drip irrigation is: **Maharashtra**
- ☆ Which type of plants is more drought tolerant: **CAM**
- ☆ Which is the stomata closing type of anti-transpirant: **Atrazine**
- ☆ Most widely adopted mechanical measures of soil and water conservation methods in hilly area is: **Bench terracing**
- ☆ Bench terracing is possible up to: **33 per cent**
- ☆ Percentage of effective rainfall under rainfed conditions is: **20-50 per cent**
- ☆ Which is the potent life saving measure under moisture stress: **Runoff recycling**
- ☆ Size of farm pond per hectare of area is: **250-300 m^2**
- ☆ Most optimum stage for supplemental irrigation from farm pond to sorghum crop is: **Flowering**
- ☆ Most economical lining materials for large pond is: **Natural Clay**
- ☆ Relationship between intensity of rainfall and soil loss is: **Exponential**
- ☆ Soil Erodability factor of normal soil is: **0.0 to 0.6 per cent**
- ☆ US Soil Conservation Service divided the soil into classes according to its capability: **8**
- ☆ Which part of India has more erosion problems: **North East India**
- ☆ Most erodible sized particles is due to wind is: **<0.1mm**
- ☆ Contour bunding is suitable in soil: **Medium soil**
- ☆ Which of the type of bench terracing is suitable for paddy cultivation in hilly areas: **Table top bench terrace**
- ☆ Broad bed and furrow is evolved in India: **ICRISAT**
- ☆ Wind erosion is major problem in: **Western India**
- ☆ Importance of fruits and vegetables in human diet is primarily because they are good source of: **Vitamins and minerals**

- ☆ Cultivation of crops in areas where rainfall is less than 750 mm per annum is termed as: **Dry farming**
- ☆ Different crops are raised in sequence or simultaneously is termed as: **Intensive cropping**
- ☆ These are arid regions with a length of growing season (period of adequate soil moisture) less than 75 days. **Dry farming**
- ☆ Moisture conservation practices are necessary for crop production: **Dry farming**
- ☆ Cultivation of crops in regions with annual rainfall more than 750 mm is in: **Dry Land farming**
- ☆ Moisture conservation practices are necessary for crop production however, adequate drainage is required especially for Vertisols is in: **Dry Land farming**
- ☆ Crop production in regions with annual rainfall more than 1150 mm, Crops are not subjected to soil moisture stress during the crop period, Emphasis is often on disposal of excess water and these are humid regions with growing period more than 120 days are the characteristics of: **Rainfed Farming**
- ☆ Areas with more than 750 mm rainfall are suitable for: **Double cropping**
- ☆ The farming system that involves the production of crops requiring tillage is: **Arable farming**
- ☆ Growing of crops in between Kharif and Rabi season is called as: **Zaid cropping**
- ☆ The grand period of rainfall in India is: **South west monsoon (SWM)**
- ☆ Alley cropping is: **Perennial hedges with crops**
- ☆ The ratio between gross sown area and net sown area is called: **Cropping Intensity**
- ☆ Pedology considered soil is a: **Natural body**
- ☆ The plants with male and female flowers on different plants are called: **Dioecious**
- ☆ Major cropping system of trans - gangetic plains is: **Rice – Wheat**
- ☆ Type of microorganism largely present in soil:**Bacteria**
- ☆ The dry matter weight accumulation in plants is due to: **Nitrogen and Carbon**
- ☆ Arrangements of soil particles is known as: **Soil structure**

- ☆ Energy metabolism in plants performed by: **Phosphorus**
- ☆ Amide form of nitrogen is in: **Urea**
- ☆ Plants obtain nitrogen from soil mostly in the form of: **Nitrate**
- ☆ *Azolla* is a symbiotic association with: **Fern and BGA**
- ☆ *Azotobactor* is useful to: **Maize, Sugarcane**
- ☆ Plants takes phosphorus in the form of: **Phosphoric acid**
- ☆ Rock phosphate is very useful to: **Acidic soil**
- ☆ Well flocculated soil is rich with: **Sodium**
- ☆ When organic matter is added to the soil: **Bulk density decreased**
- ☆ In submerged soil the decomposition of organic matter is carried out by: **Bacteria**
- ☆ The fertilizer application very close to the plant is: **Hill placement**
- ☆ The well decomposed organic substance in soil is: **Humus**
- ☆ The force of attraction that binds the molecules of different kinds is by: **Adhesion**
- ☆ The force of attraction that binds the molecules of sa kinds is by: **Cohesion**
- ☆ The flooded soil rich with hydrogen sulphide is known as: **Akiochi soil**
- ☆ The accumulation of sodium ions on the exchange sites in a soil leads to: **Alkalization**
- ☆ Fertilizer grade anhydrous ammonia contains nitrogen as**: 82 per cent**
- ☆ Production of ammonia from biological decomposition of organic nitrogen compounds: **Ammonification**
- ☆ Apatite is the naturally occurring organic complex material used for most of the phosphate fertilizer rich with: **Calcium phosphate**
- ☆ The non-volatile residue from complete burning of organic matter is: **Ash**
- ☆ The process in which photosynthates are used in building protoplasm and cell wall: **Assimilation**
- ☆ The Baul unit refers to the fertilizer required to maximize the crop yield by: **50 per cent**
- ☆ Biuret content in the commercial fertilizer should not exceed**: 0.2 per cent**
- ☆ Boron content in Borax is: **11 per cent**

- ☆ The leaching of sodium and removal of salts from the soil profile is: **Solarization**
- ☆ Biological fixation of nitrogen is helped by: **Molybdenum**
- ☆ Whiptail in cauliflower is due to the deficiency of: **Molybdenum**
- ☆ Conversion of organic form of nutrients in to inorganic form: **Mineralization**
- ☆ Biuret content in urea should be less than: **1.5**
- ☆ Shortening in Bhendi is due to the deficiency of: **Boron**
- ☆ Enriched FYM contains FYM and SSP in the ratio of**: 5: 1**
- ☆ Acid soils are characterised by a pH of: **< 6**
- ☆ Average composition of humus includes: **Carbon, Oxygen and Nitrogen**
- ☆ Absorption of water by roots in saline soils is affected by: **Osmotic effect**
- ☆ Which type of soil is commonly known as 'hungry and thirsty soil': **Red soil**
- ☆ Organic matter is abundantly available in: **Clay loam**
- ☆ A non expanding type of clay mineral is: **Montmorillionite**
- ☆ Carbon sequestration is facilitated by: **Agro Forestry**
- ☆ First scientific studies on erosion control was initiated by: **Wollny**
- ☆ Most important weather parameter influencing soil reaction is: **Rainfall**
- ☆ Which form of S increase pH: **Sulphide**
- ☆ pH range of acid rain is: **2.0-4.5**
- ☆ Optimum pH range for better survival of micro-organisms is: **6.0-7.5**
- ☆ Acidic sulphate soils are more in: **Kerala**
- ☆ Aggregate of measurement or totality of individual is called as: **Population**
- ☆ Which is the relative measure of dispersion: **Coefficient of variation**
- ☆ Which is used to know the significant difference among the treatments is real or not: **F test**
- ☆ ANOVA is useful in deviation of an individual's treatment mean from the grand mean is called as effect: **Treatment**
- ☆ In factorials experiments, 2nd denotes: **2 levels**
- ☆ Rice crop requires the plot size of for better experimental results: **20-40 m^2**

- ☆ Generally recommended formula for square root transformation for zeros and < 10 is: **X+0.5**
- ☆ Central Coastal Agricultural Research Institute (CCARI) is located at: **Goa**
- ☆ Famous rice variety "Heera" is mainly known for: **High protein**
- ☆ For which there is no national bureaus under ICAR in India: **Water**
- ☆ National Crop Insurance Programme (NCIP) is launched in: **2013**
- ☆ Cropping intensity of India is: **138 per cent**
- ☆ Rashtriya Krishi Vikas Yojana (RKVY) was launched in: **August 2007**
- ☆ Current Secretary-General of World Meteorological Organization (WMO) is: **Michel Jarraud**
- ☆ Who is Present Director-General of India Meteorological Department (IMD): **Laxman Singh Rathore**
- ☆ Which is GM crops cultivated in India: **Bt cotton**
- ☆ ICAR is a: **autonomous society**
- ☆ Energy required for evaporation: **590 cal/g**
- ☆ No. of Kishan call center: **1800-180-1551**
- ☆ Indian Grape Processing Board is located at: **Pune**
- ☆ "Rule 1989" is related with: **Genetically Modified Crops**
- ☆ National Seed Research and Training Centre (NSRTC) is located at: **Varanasi, UP**
- ☆ Ladybird beetle is: **Predator**
- ☆ Blue beetle is major pest of: **Soybean**
- ☆ First CO_2 acceptor in C-3 pathway: **RuBP carboxylase**
- ☆ From 1960's onward which of the following operation is in effect for milk: **Operation flood**
- ☆ Price fixed by government recently for agricultural products: **Minimum support price**
- ☆ Most widespread soil fertility problem in both irrigated and unirrigated areas of India is: **Phosphorus deficiency**
- ☆ Randomization reduces: **Systematic error**
- ☆ Regression coefficient lies between: **-infinite to +infinite ($-\propto$ to $\propto$)**
- ☆ Mode is: **Measure of central tendency**
- ☆ Theory of Inheritance given by: **G.J. Mondel**

- ☆ Golden rice is rich source of: **Vitamin A**
- ☆ Dr. N.E. Bourloug has been awarded with Noble Prize for: **Peace**
- ☆ "Aghanibora" is variety of: **Rice**
- ☆ Epistasis is taken from a word of: ***Greek***
- ☆ Bateson and Punnet discovered linkage in: **Sweet pea**
- ☆ Which one form 'Hartig net': **Ectomycorrhiza**
- ☆ Cell linkage theory was proposed by: **Virchow**
- ☆ Golgi apparatus is also known as: **Dictysome**
- ☆ Who discovered triangle of brassica: **Negaheru**
- ☆ Panama wilt is a disease of: **Banana**
- ☆ Jute is mainly grown in: **Eastern part of India**
- ☆ The term Evergreen Revolution was given by: **M.S. Swaminathan**
- ☆ The average concentration of carbon dioxide in the atmosphere is: **300 ppm**
- ☆ Agrostology is the study of: **Grasses**
- ☆ Atmospheric layer nearest to earth Surface is: **Troposphere**
- ☆ Plant that grow on extremely dry soil are classified under: **Xerophytes**
- ☆ Ground water table is measured by: **Piezometer**
- ☆ Mycorrhiza is associated with what part of plants: **Roots**
- ☆ The most Suited N fertilizer for tea is: **Ammonium sulphate**
- ☆ Which food is designated as "Boneless meat": **Soybean**
- ☆ Which of the following control the root initiation, cell elongation and apical Dominance: **Auxins**
- ☆ Study of soil from the stand point of higher plant is known as: **Edaphology**
- ☆ Pruning is most essential for: **Tea**
- ☆ The progeny of breeder seed is: **Foundation seed**
- ☆ The basis of farm budgeting is: **Cost – benefit analysis**
- ☆ The largest tea production state in India is: **Assam**
- ☆ The plant nutrient which help in translocation of sugars and starch is: **Potassium**
- ☆ Saffron is largely produced in: **Jammu and Kashmir**

- ☆ A crop that can supply oil for petrochemicals is: **Jatropha**
- ☆ Most widely cultivated mustard type in India is: **Indian mustard**
- ☆ C_4 plant normaly produce more biological yield than C_3 plant because of: **Less photorespiration**
- ☆ Absolute zero is: **273 ?K**
- ☆ In India maximum area under wheat is occupied by the Species: ***Triticum aestivum***
- ☆ Which among the following coloures absorbs more radiation: **Black**
- ☆ Which of the following elements is most mobile in soil: **N**
- ☆ The most deficient micro – nutrient in Indian soil is: **Zn**
- ☆ Magnesium is a constituent of: Chlorophyll
- ☆ The C: N ratio of arable soil commonly ranges from:**8: 1 to 15:1**
- ☆ Dwarfing gene in wheat is: **Norin**
- ☆ Which of the following element immobile in soil but mobile in plants: **P**
- ☆ Which soil has highest efficiency: **Clay soil**
- ☆ Dangerous gas for depletion of ozone layer is: **CFC**
- ☆ Which of the following is not a biopesticide: **Carbaryl**
- ☆ Major cropping system of trans-gangatic plains is: **Rice –wheat**
- ☆ Photosynthetic inhibition by O_2 is called: **Warburg's effect**
- ☆ Crop logging is done in: **Sugarcane**
- ☆ Blue revolution is related with: **Fish**
- ☆ Inflorescence in rice is known as: **Panicle**
- ☆ Alluvial soils are found in: **River delta**
- ☆ Weight of one cotton bale is: **170 kg**
- ☆ Hybrid cotton in India was evolved for the first time in: **1970**
- ☆ The relative proportion of sand, silt and clay is called: **Soil texture**
- ☆ Which of the following crops is thermo-insensitive: **Sunflower**
- ☆ The IARI was established in: **1905**
- ☆ In which of the following crops GM varieties are available for cultivation in India: **Cotton**
- ☆ Photo-periodically rice is a: **Short day plant**

- ☆ The present level of carbon-dioxide in atmosphere is: **295 – 300 ppm**
- ☆ Maize belongs to the category: **Monoecious**
- ☆ Pheromone trap attracts: **Male moths**
- ☆ Origin place of soybean is: **China**
- ☆ India rank first in the production of the following crops in the world: **Pigeonpea**
- ☆ In plants, enzyme responsible for the synthesis of the malic acid is: **Rubisco**
- ☆ Which soil has highest efficiency: **Clay soil**
- ☆ Which of the following insecticides may be recommended for the control of Termites: **Chlorpyriphos**
- ☆ Correct order of rice producing countries is: **China> India> Indonesia> Thailand**
- ☆ The depth of seeding in wheat is depends on: **Length of coleoptiles**
- ☆ IGFRI is located at: **Jhansi**
- ☆ The photo synthetically active (PAR) falls in the range of: **400 – 700 nm**
- ☆ Missing data are calculated by using: **Missing plot technique**
- ☆ Soil mulch is useful in: **Minimize evaporation losses**
- ☆ PAR (Photo-synthetically active radiation) is measured in: **Einstein**
- ☆ The study of relationship between properties and plant production is as: **Edaphology**
- ☆ Triticale is a cross between: **Wheat x Rye**
- ☆ Maximum number of treatment accommodates in RBD without loss of efficiency is: **20**
- ☆ Photo-respiration rate is highest in which group of plants: **C3 plants**
- ☆ The first maize hybrid in India was: **Ganga-l**
- ☆ Aflatoxin contamination generally found in: **Groundnut**
- ☆ Which crop is also known as white gold: **Cotton**
- ☆ Required seed rate for raising tomato nursery is: **400 gm**
- ☆ Test between two population variance is done by: **F – test**
- ☆ Which test is used for comparing two means from independent samples: Z- test

- ☆ CO_2 accepter in C4 plants is: **OAA**
- ☆ The chemical responsible for lathyrism in mammals is: **BOAA**
- ☆ Which element is involved in bio- synthesis of IAA: **Zinc**
- ☆ The area under Bt-cotton in India is about: **6.4 mha**
- ☆ The net gain of ATP in glycolysis is: **2 ATPs**
- ☆ The precursor of ethylene is: **Methionine**
- ☆ Which state contributes maximum pulse production: **M.P.**
- ☆ Which of the following crops has the largest area under transgenic globally: **Soybean**
- ☆ Sugar-beet is an indicator plant for: **Molybdenum**
- ☆ In India, area under rice is about: **45 Mha**
- ☆ Sunflower is also known as an individual plant for the deficiency of: **Boron**
- ☆ When the fertility gradient of the field is in two directions, the most appropriate
- ☆ Experimental design is: **LSD**
- ☆ Plants capable of growing in rocks crevices are called: **Lithophytes**
- ☆ The C: N ratio of humus is: **10: 1**
- ☆ Densest part of atmosphere strata is: **Troposphere**
- ☆ Agroclimatic zones in India are**: 15**
- ☆ Law of Minimum was given by**: Liebig**
- ☆ Sulphur containing amino acid: **Methionine**
- ☆ The specific heat of water is: **1 Cal/g**
- ☆ Disease caused by Zn deficiency is: **Khaira**
- ☆ Mat nursery is related to: **Rice**
- ☆ ANOVA' was given by: **R.A. Fisher**
- ☆ Karnal bunt is caused by: ***Nevosia indica***
- ☆ Bonneville, Early Badger, Arkel are the improved varieties of: **Soybean**
- ☆ Most of the wheat varieties are: **Short day plants**
- ☆ The simplest measure of variability in a data set is: **Range**
- ☆ The crop having the highest pesticide use is: **Cotton**

- ☆ Double cross hybrid maize production technique was introduced by: **D.F.Jones**
- ☆ 'Vertisol' is related to: **Black soil**
- ☆ The square of standard deviation is: **Variance**
- ☆ The principle of making use of greater homogeneity in groups of experimental units reduce the experimental error is: **Local control**
- ☆ Sugarcane inflorescence is: **arrow**
- ☆ Country having the maximum area of hybrid rice is: **China**
- ☆ Rice grain is deficient in: **Lysine**
- ☆ An ideal type of rice with small, thick and erect leaf was proposed by: **Yoshida**
- ☆ Biurate content in urea is: **1.5 per cent**
- ☆ Maximum productivity of sugarcane in India is: **Tamil Nadu**
- ☆ Non-edible plant suitable for biodiesel: **Jatropa**
- ☆ Saffron (kesar) belong to the family**: Iridaceae**
- ☆ The 5- carbon compounds produced during dark reaction of photosynthesis is: **Ribulose bis phosphate**
- ☆ Most suitable design for experiment involving varying number of tillage and nitrogen levels is: **Split plot**
- ☆ In an RBD experiment having 9 treatments and 4 replications, the error degree of freedom will be: **24**
- ☆ The plants with male and female flowers on different plants are called: **Dioecious**
- ☆ Which one of the following process results in release of energy: **Catabolism**
- ☆ Which one of the following element is a constituent of protoplasm: **Calcium**
- ☆ Which one of the following element is mobile in plants but immobile in soil: **Phosphorus**
- ☆ Which of the following process is not takes place in evolution of plants: **Linkage**
- ☆ Solar constant is equal to (in cal/cm^2/min): **1.94**
- ☆ Which one of the following is sulphur-containing amino acid: **Cystine**

- Horse latitude lies in between north and south latitudinal belt of equator: **30 to 35**
- The splitting up of water molecules in plant cells in the presence of sunlight is called: **Photolysis**
- Norin – dwarf gene was introduced in which cereal: **Wheat**
- The centre of origin of *Triticum aestivum* is: **Mediterranean centre**
- Which state of India leads in area under wheat crop: **UP**
- The process of use of microorganism to remove salts from soil is referred as: **Bioremediation**
- Economic part of Isabgol is: **Seed and husk**
- Which part of the plant cell is known as power house: **Mitochondria**
- Which one of the following clay mineral has the highest CEC: **Vermiculite**
- The light generated reducing power is: **$NADPH_2$**
- *Triticum aestivum* is: **Hexaploid**
- Gregor John Mendal worked on the crop**: Garden pea**
- t-test is applicable when the numbers of treatments are**: 12**
- The first Director General of ICAR**: Dr. B. P. Paul**
- Red and purple colour of maize (***Zea mays***) is due to deficiency of: **P**
- Most common method of irrigation in India is**: Check basin irrigation**
- Absorption of solute ions is regulated by: **Vacuoles**
- One gram of glucose produces how much fat or oil: 0.45 g
- Growth of plant towards light is called**: Photoperiodism**
- The plants which open their stomata during night for taking CO_4 are known as: **CAM**
- Plants absorb phosphorus in the form of: **$H_3PO_4^-$**
- Principle of **Experimental design** is given by**: R.A.Fisher**
- An acid derived from green leaves of chickpea prescribed for **intestinal** disorders is**: Oxalic acid**
- Atmosphere extends above mean sea level to a height of about: **1600 km**
- The Oleoresin compounds occurs in: **Chilli**
- The law of tolerance was introduced by: **Shelford**

- End product of glycolysis is: **Pyruvate**
- Plant cells are connected with the help of: **Plasmodesmata**
- Which one of the following is not a legume crop: **Wheat**
- The two major races of rice are: **Indica and Japonica**
- Which of the following state has the largest geographical area: **U.P.**
- The highest production in terms of million tonnes per year in India is that of: **Pulses**
- Respiration in plant is essentially a process related to the following: **Oxidation**
- Linkage between gene affects: **Anaphase**
- DNA contains following number of nitrogenous base: **Four**
- Animal and plant cell can be differentiated by: **Presence or absence of cell wall**
- In temperate countries, sugar is generally obtained from which of the following crops: **sugar beet**
- The idea of particulate nature of inheritance was given by: **Darwin**
- Alternate form of gene at the same locus are referred to as: **Allele**
- Sequence of growing crops in a piece of land is known as: **crop rotation**
- In diploid species generally a following number of the chromosomes are involved in the pollen mitosis: **4n**
- Sulphur fungicides can be freely used on all vegetable except the following: **Cucurbits**
- Karnal bunt is a serious disease of: **Wheat**
- Zinc is required for the synthesis of: **Tryptophane**
- Contour system of orchard planting is generally followed in: **Hills**
- NARS refers to: **National Agricultural Research System**
- Cry-preservation is done in liquid nitrogen at a temperature of: **-196°C**
- Crossing over during meiosis results in: **Breaking linkage**
- The simplest measure of variability in data set is: **Range**
- Study of fresh water is called: **Hydrology**
- The GM crops has highest area in order of preference: **Insecticide then herbicide resistance**

- ☆ "AADHAAR" indicates: **Unique Identification card**
- ☆ Artificial bacteria culture is developed by: **Dr. Jenial James**
- ☆ Blue tag in insecticide is for indication of: **Slight toxicity**
- ☆ Which one is Indian origin fruit crop: **Ber**
- ☆ For virus free seed use: **Meristem tip** culture
- ☆ Which variety of rice is known for longest seed length: **Pusa-1121**
- ☆ Which one is neutral fertilizer: **CAN**
- ☆ 12. ELISA test is used to detecting: **Virus**
- ☆ 13. Common bee species cultivated worldwide is: ***Apis dorsata***
- ☆ Which organelle is involved in root gravity pool: **Amyloplast**
- ☆ White bud of maize is due to: **Zn**
- ☆ Mobility stage of seed is due to: **ABA**
- ☆ Which one is not measure of central tendency: **Mean deviation**
- ☆ Which one is not measure of dispersion: **Range**
- ☆ Ear cockle disease is due to: ***Anguina tritici***
- ☆ Support price fixed on the recommendation of
- ☆ The correlation coefficient will be negative when: **x increase and y decrease**
- ☆ *Phalaris minor* of wheat crop controlled by: **2, 4-D**
- ☆ Crops Grown on between two main crops is called: **Intercrops**
- ☆ Which one is the source variety of resistant for stem borer, BLB and Blast of rice: **IR 36**
- ☆ Vitamin B_1 is also known as: **Thiamine**
- ☆ IVLP stands for: **Institute Village Linkage Programme**
- ☆ Which one of these is major constituent of poultry feed: **Maize**
- ☆ In binomial distribution: **Mean > variance**
- ☆ Bordeaux mixture is: **Fungicide**
- ☆ Which of the following is used to turn over the soil: **Mould Bould plough**
- ☆ NBFGR-National Bureau for Fish Genetic Resource is situated at: **Lucknow**
- ☆ Which of the following is complex fertilizer: **Urea ammonium Phosphate**

- ☆ Soil having ESP (Exchangeable Sodium Percentage) greater than 15 are: **Alkali soil**
- ☆ Rain, mist, fog and cloud all these phenomena occurs in: **Troposphere**
- ☆ Which nutrient helps in Biological Nitrogen Fixation: **Mo**
- ☆ CIMMYT works on: **Maize and wheat**
- ☆ Which of the following is highly salt tolerant fruit crop: **Date palm**
- ☆ Disease occurring regularly in the same area is called as: **Endemic**
- ☆ Which of the following is essential component of nucleic acid and protein: **Nitrogen**
- ☆ Banana is: **Auto triploid**
- ☆ Which crop in India has maximum area under irrigation: **Wheat**
- ☆ Which of the following is not found in plant cell: **Glycogen**
- ☆ Family of cotton is: **Malvaceae**
- ☆ Certified seed is produced from: **Foundation seed**
- ☆ In prophase which is correct: **Elongated threads like chromosome**
- ☆ In cereals, grain starch is consisting of: **Amylopectin and Amylase**
- ☆ The National Agricultural Insurance Scheme was started in: **1999-2000**
- ☆ Flaver Savr is a variety of: **Tomato**
- ☆ Cotton leaf curl virus is transmitted by: **White fly**
- ☆ The region in India where boron deficiency is mostly occurs: **North Western region**
- ☆ The brown box dots on food package indicates: **Organically produce product**
- ☆ Land resources in India as a whole world is: **2 per cent**
- ☆ The soil order vertisols is related to: **Black soil**
- ☆ Which state has a maximum area in fruit production: **Maharashtra**
- ☆ The statistical test used to determine the goodness of fit is: **Chi- square test**
- ☆ The cardigan quarto protocol for agriculture product was came in the year of: **2005**
- ☆ The citrus canker disease of lemon was introduced in India from: **USA**
- ☆ Utera cultivation is practiced mainly in: **Chhattisgarh**

- ☆ In the pulse crops, which fertilizer is advised to basal application: **SSP**
- ☆ The soil amendment mainly used for reclamation of sodic (alkali) soil is: **Gypsum**
- ☆ The biogas has major content with CO_2 is: **CH_4**
- ☆ The most common method of hybridization for wheat is: **Gogo method**
- ☆ Hatch and slack pathway is found in: **Maize**
- ☆ The most important process in chemical weathering: **Hydrolysis**
- ☆ Vector of leaf curl disease of cotton and tobacco: ***Bemesia tabaci***
- ☆ Vector of tomato spotted wilt: **Thrip**
- ☆ Reagent used for the extraction of phosphorus in alkaline soils: **Olesen's reagent**
- ☆ Pusa delicious is type of papaya variety: **Gynecious**
- ☆ The first insect resistant variety 'under hill' is related to: **Wheat**
- ☆ Odomos the mosquito repellent contains: **Pyrithrin**
- ☆ Based on Severity of damage Wheat pests which have been ranked in the order: **Brown rust >Loss smut>Termites>Karnal bunt**
- ☆ India has monopoly in the production of: **Muga silk**
- ☆ The most common form of DNA is: **B DNA**
- ☆ Replication of DNA is: **Semi conservative**
- ☆ Which is the first cotton hybrid: **H4**
- ☆ Nationalization of banks was in the year: **1969**
- ☆ Livestock loan is a: **Medium term**
- ☆ Correlation co-efficient is independent of**: Origin and Scale**
- ☆ Accept when null hypothesis is false: **Type II error**
- ☆ Richmond Lang effect is associated with: **Cytokine**
- ☆ The soil pH range in which most of the nutrients are available: **6.5-7.5**
- ☆ Pahala blight of sugarcane is due to the deficiency of: **Mn**
- ☆ Red soils mainly come under the order: **Alfisol**
- ☆ Central Agmark Laboratory is located at: **Nagpur**
- ☆ Buffer stock is maintained by: **FCI**
- ☆ Who discovered Bordaux Mixture: **PMA Millardet**

- ☆ First discovered bacterial disease of plants: **Fire blight**
- ☆ Alternate host of *P. graminis triticii* is: **Beriberi's**
- ☆ Organic matter generally contains of organic carbon: **1.724 per cent**
- ☆ TRYSEM was introduced in the year: **1979**
- ☆ High yielding variety program was launched in: **1965**
- ☆ The Chairman of national development of council in India: **Prime Minister**
- ☆ NARP was launched by: **Govt. of India in 1979**
- ☆ Largest state in India: Rajasth**an**
- ☆ Father of hybrid rice: **Y.L. Ping**
- ☆ Famous Indian pathologist, extensively studied about cereal rust: **K. C. Mehta**
- ☆ Grey speck of oat is due to the deficiency of: **Mn**
- ☆ Prominent foul smell of Karnal bunt is due to: **Trimethyl amine**
- ☆ A phenomenon wherein a single gene lies more than one phenotypic effect is known as: **Pleiotropism**
- ☆ Which one of the following is a rich source of protein: ***Spirulina***
- ☆ Glycolysis is also known as: **EMP Pathway**
- ☆ DNA and RNA are similar to each other because both are: **Nucleotide polymers**
- ☆ A simple technique has been developed in India by using the: ***Azotobacter* and *Rhizobium***
- ☆ The insects excrete in the form of uric acid because: **most of them are terrestrial**
- ☆ Measurement of the rate of oxygen consumption in unit volume of water over a period of time is carried out to determine: **Biological Oxygen Demand**
- ☆ Biosphere is made up of: **organism, lithosphere, atmosphere and hydrosphere**
- ☆ Red data book is famous for: **Endangered animals and plants**
- ☆ Bottom layer of deep fresh water habitat is: **Hypolimnion**
- ☆ International Conversion on Biological Diversity was ratified in the year: **1994**
- ☆ Which is considered as biological paradise in India: **Nilgiris Bioreserve**

- ☆ Which of these is often called as "cell organizer": **Nucleous**
- ☆ Hardness of woody tissue is due to: **Lignin**
- ☆ Pea seeds will germinate best if soaked in ajar containing: **O_2**
- ☆ The Anticodon region is present in: **hn- RNA**
- ☆ The reptiles and birds, nitrogen is excreted in the form of: **Lipids**
- ☆ Red orange colour in saffron is due to: **Crocetin**
- ☆ Beaten rice is known as: **Flaked rice**
- ☆ Roasting coffee beans improves the content of: **Niacin**
- ☆ The crop grown for grain, green manure and fodder: **Chick pea**
- ☆ Black tip of mango – **B deficiency**
- ☆ Half life of P_{32} is **14 days**
- ☆ Calcium requirement highest in **legumes**.
- ☆ Maximum CEC at root surface noticed in **legumes**.
- ☆ Maximum infiltration in soil at bulk density **1.1**
- ☆ At saturated condition in soil matric potential is **zero**.
- ☆ Mn toxicity is found in **Assam**.
- ☆ Deposition of material with the action of gravitational force is called **colluvium.**
- ☆ As the soil changes from acid to alkali redox potential: **decreases**.
- ☆ Fools gold: **pyrite**
- ☆ First world food prize was awarded to **Dr. M.S.Swaminathan**, (1987) for introducing high yielding wheat and rice varieties to India starting Indian green revolution.
- ☆ 1 kpa:**10 mb**
- ☆ 1° latitude: **111 km**
- ☆ Thermometer discovered by: **Tricolle.**
- ☆ Epigeal germination – **green gram, bean, castor, mustard, tamarind, sunflower, papaya, cotton, gourd, onion, guar, cucumber.**
- ☆ Hypogeal germination – **chickpea, groundnut, wheat, pea, mango, monocotyledons, arecanut, coconut, Runner bean (*Phaseolus coccineus*)**
- ☆ In remote sensing optical wavelength used is: **0.3 – 15μm**.

- ☆ Biodiversity act: **2002**
- ☆ Insecticide act – **1968**
- ☆ Cellulose content in humus – **21 per cent**
- ☆ Nitrogen content in humus – **5 to 5.5 per cent**
- ☆ Primary electron acceptor in **photosystem II is pheophytin**
- ☆ Recommended safe limit of land slope for border irrigation in medium soils is **0.2 – 0.4 per cent**
- ☆ Element required for stabilizing various enzyme systems in plants is **phosphorus**
- ☆ Plant quarantine order passed in **2003**
- ☆ **Square root transformation** is done when observations follow poison distribution.
- ☆ Movement of water molecules from a solution with higher concentration towards lower concentration solution through semi permeable membrane is called **osmosis.**
- ☆ Gypsum equivalent of iron pyrite is **0.63**.
- ☆ Most suitable system for arid regions is **pasture development**.
- ☆ Agroforestry system suited in low rainfall areas is **silvipasture.**
- ☆ Earth emits **long wave radiation**.
- ☆ **Albedo** is highest for **fresh snow**.
- ☆ CO_2 compensation point in C_4 plants is **0-10 ppm.**
- ☆ CO_2 compensation point in C_3 plants is **30-70 ppm.**
- ☆ Photosynthetically active radiation – **400 to 700 nm**
- ☆ Climate where low temperature in winter and high temperature in summer prevails is **Subtropical climate.**
- ☆ Largest canal for the purpose of irrigation in world is Indira- **Gandhi canal.**
- ☆ In India contribution of herbicides in pesticide consumption is **16 per cent**
- ☆ Protection of plant varieties and farmer rights authority – **New Delhi**
- ☆ Anti-gibberellins in action - **Abscisic acid**
- ☆ Nitrate is commonly absorbed by **Phloem**
- ☆ End cytokinin is derivative of **Adenine (purine)**

- ☆ Balanced fertilizer ratio for legumes is **1:2:2**
- ☆ Blue colour triangle on insecticide container represents **slightly toxic.**
- ☆ Mechanical analysis of soil makes use of **stokes law.**
- ☆ First KVK at Pondicherry started in **1974.**
- ☆ Humic acid which is a fraction of humus is **Soluble in alkali, insoluble in acid.**
- ☆ Average content of Si in soil – **27 per cent**
- ☆ Fulvic acid → Humic acid → Recalcitrant biopolymers → **humin acid**
- ☆ Humic acid is aromatic with **high molecular weight**
- ☆ Humus fraction lightest in colour is **Fulvic acid**
- ☆ Humus forming fungi includes the genera ***Alternaria* and *Aspergillus*.**
- ☆ Fulvic acid is soluble **both in acid and alkali.**
- ☆ Calibrated device used for measuring flow of water in open conduit is **Parshall flume.**
- ☆ National agriculture policy aims the growth rate of **> 4 per cent per annum.**
- ☆ Term evergreen revolution given by **M.S. Swaminathan.**
- ☆ Phytohormone involved in nutrient mobilization is **cytokinin.**
- ☆ Floral abnormality is present in white rust of **crucifers.**
- ☆ Floral abnormality absents in downy mildew of **mustard.**
- ☆ Illite is rich in **potash.**
- ☆ Antisenescence phytohormone is represented by **gibberellins**
- ☆ **Al_3^+** has higher aggregation capacity in **soil.**
- ☆ **Na^+** has low adsorption capacity on clay compared to H^+, Ca^{2+}, Fe^{2+}.
- ☆ Slow growing species of legume root nodule bacteria included in the bacterial genus ***Bradyrhizobium.***
- ☆ Seed work board is required for **seed blending.**
- ☆ Soil structure improved by application of **superphosphate.**
- ☆ Approximate weight of 15 cm surface soil over 1 ha is **2.24×10^6 kg.**
- ☆ Paddy causes maximum reduction in soil **alkalinity.**
- ☆ Flow of water in saturated soil is described by **Darcy's law and Poiseuille's law.**

- ☆ True density of most mineral soils is **2.65 g/cm^3.**
- ☆ Rainfed authority of India – **New Delhi.**
- ☆ Maximum K fixation takes place in **Vermiculite.**
- ☆ Fast growing tree species – ***Eucalyptus.***
- ☆ ***Albizzia lebbek*** – multipurpose trees species
- ☆ Preventing fruit drop in grapes by spraying **NAA @ 20 ppm.**
- ☆ **Casparian strip** is present in endodermis.
- ☆ Most effective instrument for measuring soil moisture content insitu is **TDR**.
- ☆ Freezing point depression of a chemical solution is measured by **osmomete**r.
- ☆ Average human labour power in India is **0.1 KW**.
- ☆ Maximum area under fruit cultivation in the state – **Maharashtra**
- ☆ Transgenic crop having maximum cultivated area in the world is **Soyabean.**
- ☆ In cereals grain starch consists of **amylase and amylopectin**.
- ☆ Highest Zn containing fertilizer is **ZnO**.
- ☆ Apical dominance in sugarcane controlled by **cytokinins**.
- ☆ Water soluble vitamins – **B, C**
- ☆ Fat soluble vitamins – **A, D, E, K**
- ☆ According to area requirement **Basin > Catchment > Watershed**
- ☆ Monosaccharides – **Glucose, Galactose, Ribulose**
- ☆ Disaccharides – **Maltose, Lactose, Sucrose, Cellobiose**
- ☆ Non reducing sugar – **sucrose**
- ☆ Sugar largely found in germinating seeds – **Maltose**
- ☆ Glucose + Galactose = **Lactos**e
- ☆ **Glycogen** is present only in animal cells
- ☆ Positively charged aminoacid is **lysine**
- ☆ Sugar sweetest among all sugars is **fructose**
- ☆ Enzyme for conversion of ammonia to amino acid is **glutamine synthetase spherosomes** are related with fat

- ☆ Ribosomes are produced in **Nucleolus**
- ☆ Pigments that produce the colour of many flowers or red of red maple leaves are stored in **vacuoles.**
- ☆ Ultimate electron donor in mitochondrial electron transport chain is **ubiquinone.**
- ☆ Latent heat of vaporization (water to vapor) of water is **586 cal**.
- ☆ Latent heat of fusion (ice to water) is **80 cal.**
- ☆ Important radiant generated in Pentose phosphate pathway of glucose degradation is **NADPH.**
- ☆ **Glyoxysomes** function in breakdown of fatty acids.
- ☆ Maximum free radical's production takes place in **senescence**.
- ☆ Precursor for ethylene biosynthesis is **methionine**.
- ☆ In a rapidly transpiring plant, the water column in xylem will be under **negative pressure**.
- ☆ Transpiration is measured by **potometer**.
- ☆ Proteins that bind to tata box in promoter's region are **transcriptional factors**.
- ☆ Chemical nature of GA_3 is **terpene**.
- ☆ Site of oxidative electron transport in cell is **mitochondria**.
- ☆ Guttation is not favored **under low root pressure.**
- ☆ Xylem and phloem elements in plant are surrounded by a layer of living cells called **pericycle.**
- ☆ Part of root that absorbs water and minerals is **root hairs.**
- ☆ Stage of seeds showing no germination because of internal conditions of seed is known as **dormancy.**
- ☆ Plant hormone which is primary regulator of abscission process is **ABA.**
- ☆ Most of the wheat cultivars are qualitative **long day plants.**
- ☆ Rice grain is deficient in **lysine**.
- ☆ Critical concentration of micro-nutrients needed in tissue is equal to or **less than 100ppm.**
- ☆ Storage of elements in vacuoles occurs under **luxury consumption.**
- ☆ **Chelates** are organic in nature.

- ☆ Siderophores produced by bacteria and fungi in the soil are source of **iron.**
- ☆ Iron stored in chloroplast as iron protein complex called **phytoferritin.**
- ☆ Initiation of protein synthesis in eukaryotic mRNA requires a **5' cap.**
- ☆ The first sign of switchover from vegetative stage to reproductive stage in wheat is **double ridge stage.**
- ☆ Abundant phosphorus in plants is related with delayed maturity and accumulation of **anthocyanin pigments**.
- ☆ Increase in temperature at anthesis stage in wheat results in **decreased grain size**.
- ☆ Ideal type of rice with small, thick and erect leaf was proposed by **Yoshida.**
- ☆ Increase in wheat yield potential so far results from increase in **harvest index.**
- ☆ Most commonly grown crop plants are included in **glycophytes.**
- ☆ **Acetylene reduction** to ethylene is measured as nitrogen fixation.
- ☆ **Mn deficiency in crops:**
  - ❒ Grey speck of oats
  - ❒ Marsh spot of peas
  - ❒ Speckled yellows of sugarbeet
  - ❒ Pahalablight of sugarcane
- ☆ **Boron deficiency in crops:**
  - ❒ Heart rot of sugarbeet
  - ❒ Drought spot of apple
  - ❒ Stem crack in celery
  - ❒ Water core in Turnip
- ☆ **Zinc deficiency in crops:**
  - ❒ Little leaf of cotton
  - ❒ Rosette of apple
  - ❒ White bud of Maize
- ☆ Under excessive light level the synthesis of Zeaxanthin increases.
- ☆ Dieback disease in cotton is the result of **copper deficiency.**
- ☆ For the synthesis of auxins micro-nutrient essential is **Zinc.**
- ☆ Direct reduction of O_2 by photosystem I leads to the formation **of H_2O_2.**
- ☆ Optimum pH of nutrient solution in nutrient solution culture is **6.**

- ☆ Acid rains are due to **NO_2 and SO_2.**
- ☆ Association of fungi with roots of higher plants – **mycorrhiza.**
- ☆ Arbuscular Mycorrhizae (AM) mostly used in **perennials.**
- ☆ If the accumulation ratio in absorption of nutrients is greater than one then it is **active absorption.**
- ☆ Most harmful pollutant by automobiles is **carbon monoxide.**
- ☆ C_{14} has a half life of **2530 years.**
- ☆ Mass flow mechanism proposed by **Munch.**
- ☆ Isotopes differ in number of **neutrons.**
- ☆ The process in which sugars (carbohydrates) are raised to high concentration in phloem cells close to a source is known as **phloem loading.**
- ☆ Bundle sheath chloroplast: **Lack grana**
- ☆ High auxin and low cytokine favour: **Root**
- ☆ Precursor of ethylene: **Methionine**
- ☆ An international pest: ***Helicoverpa* spp.**
- ☆ Carbohydrate factory of cell is: **Chloroplast**
- ☆ In photorespiration NH_3 evolution take place at: **Mitochondria**
- ☆ Organelles involved in photorespiration: **Chloroplast, Mitochondria, Peroxisome**
- ☆ Particle size of colloids is (in microns): **1-100**
- ☆ Term genetics was coined by: **Batson**
- ☆ Components of prokaryotic ribosome are: **50s and 30s**
- ☆ Plasma membrane is: **Selectively permeable**
- ☆ ABO blood group is an example of: **Multiple alleles**
- ☆ Catabolism is the breakdown of large molecules to small molecules and this process often **releases energy.**
- ☆ Climacteric rise in respiration is observed in **mango fruit.**
- ☆ Callus is induced to form roots in the medium of more auxins than **cytokinin.**
- ☆ Sulphur containing aminoacids are **methionine and cysteine.**
- ☆ Group of enzymes which form double bond by elimination of a chemical group is **Lyases.**

- ☆ Electrophoresis was developed to **separate proteins.**
- ☆ Ripening is delayed by the synthesis of antisense ACC synthetase RNA in **tomato.**
- ☆ Blue light is always less efficient in photosynthesis than **red light.**
- ☆ Recognition site of t-RNA is **anticodon.**
- ☆ Chlorophyll is green because it reflects **green colour.**
- ☆ Plastocyanin protein contains **copper.**
- ☆ Z-scheme of electron transport was first proposed by **Hill and Bendall.**
- ☆ Number of photons required to produce 1 molecule of oxygen is **8**
- ☆ In C_3 plants RUBISCO first reacts with 2 to form **PGA.** Hormone associated with acid growth theory is **auxin.**
- ☆ Nitrate reductase is found in **cytoplasm**.
- ☆ Natural inhibitor of IAA oxidases is **coumaric acid.**
- ☆ In C_4 plants first stable product of photosynthesis is **oxalic acid.**
- ☆ Close association of chloroplast, peroxisomes and mitochondria in a leaf cell is associated with **photorespiration.**
- ☆ Number of ATP required to produce 1 mole of hexose in photosynthesis is **18.**
- ☆ Instrument used for measuring stomatal pressure is **porometer.**
- ☆ Scientist related to Biological clock – **Bunning**
- ☆ Most dangerous gas for depletion of ozone layer is **CFC.**
- ☆ Photorespiration is also known as **glycolate pathway.**
- ☆ In C_4 plants enzyme responsible for synthesis of malic acid is **PEP carboxylase.**
- ☆ Element required for dehydrogenases is **Zinc.**
- ☆ Most striking feature of CAM plants is formation of **malic acid at night**.
- ☆ **Tropical rain forest** has highest net primary productivity per unit area.
- ☆ The irradiance at which photosynthesis is equal to respiration rate (Net CO_2 exchange is zero) is **light compensation point**.
- ☆ CO_2 concentration at which photosynthetic fixation just balances respiratory loss is CO_2 **compensation point.**
- ☆ Transpiration ratio is highest for **C_3 plants.**

- End product of glycolysis is **pyruvic acid.**
- Glycolysis takes place in **cytoplasm.**
- Kerbs cycle produces **30 ATP.**
- First time IAA from human urine was isolated by **Kogl.**
- Scientists related to skototropism: **Strong and Ray.**
- Letham isolated zeatin from **corn seed.**
- Phaseoline is the storage protein in **Beans.**
- Main organic acid in pineapple is **malic acid.**
- Sulphate reduction in leaves takes place in **chloroplast.**
- Polymer of cellulose is **β-D glucose.**
- In C_4 plants PEP carboxylation takes place in **mesophyll cells.**
- Cyanide resistant respiration follows pentose **phosphate pathway.**
- Aminoacid produced in photorespiration is **serine.**
- In monocots and dicots accumulation of ethylene causes collapse and lysis of mature cortical cells in root and formation of tissue with large air spaces.
- Aerenchyma is related with **ethylene.**
- When starch reacts with iodine produces **blue colour.**
- Under anaerobic conditions microbes grow slower but utilizes more sugar and produces more CO_2 and ethanol, this phenomenon is known as **Pasteur effect.**
- Enzyme used to cut double stranded RNA is **restriction endonuclease.**
- H^+ cannot pass across membrane by **diffusion.**
- In many species gradual decrease in respiration is reversed by sharp increase known as **climacteric.**
- Conversion of organic nitrogen to NH_4 by soil microbes is called **ammonification.**
- **Denitification** occurs in waterlogged soils.
- C_3 cycle of carbon fixation takes place in stroma of **chloroplast.**
- The process by which N_2 is reduced to NH_4^+ is called **nitrogen fixation.**
- Electrons required for conversion of NO_2^- to NH_4^+ are **6.**
- Mature root nodule made largely of **tetraploid cells.**

- ☆ Nitrogen fixation is carried out by the enzyme **Nitrogenase.**
- ☆ Nitrogenase consists of **Fe-Mo protein.**
- ☆ Dormin is coined by **Wareing.**
- ☆ Kranz type of anatomy found in **sorghum (C_4 plant).**
- ☆ 6 calvin cycles needed to produce **1 molecule of glucose.**
- ☆ **Ureides** (Allantoin and allantoic acids) are the major nitrogen
- ☆ Compounds transported from root nodules to other parts of plant in **soybean.**
- ☆ Number of quanta in 1 micro Einstein is **6.02 x 1017.**
- ☆ ***Pascal*** is the SI unit of pressure.
- ☆ In the process of nitrate reduction oxidation number of nitrogen changes from **+5 to -3.**
- ☆ Reduction of nitrite to ammonium catalysed by nitrite reductase in **chloroplast and proplastids of roots.**
- ☆ Flowering stimulus is perceived by **leaves.**
- ☆ 1-amino cyclopropane 1-carboxylic acid (ACC) is a close precursor of **ethylene.**
- ☆ First step of sulphate assimilation is catalysed by **ATP sulfurylase**.
- ☆ Coconut fat is a rich source of **lauric acid.**
- ☆ Antimicrobial compounds synthesized by plants when infected with microbes are called **hytoalexins. Phytoalexins – Pisatin, Phaseolin, Isocoumarin**.
- ☆ In groundnut, low light intensity during peak flowering reduced number of **flowers per plant.**
- ☆ Flowers opened during cloudy period **do not produce pegs.**
- ☆ Low light intensity at pegging and pod filling reduces **peg and pod number.**
- ☆ In cereals, number of tillers increase with increase in **light intensity.**
- ☆ Low light intensity during flowering increases spikelet sterility in **cereals.**
- ☆ The low light intensity from panicle initiation to grain formation is critical but flowering to grain formation is more critical.
- ☆ Low light during ripening reduces yield due to lesser number of filled grains per panicle and lower grain number.

- Reduction in grain yield of rice in wet season compared to dry season is attributed to solar radiation.
- Dry matter production increases in proportion to the amount of intercepted radiation; but yield may or may not.
- In northern latitude, day length starts increasing from January onwards and reaches a peak in June and subsequently decreases up to December.
- The photoperiodic effect is realized only when optimum temperature prevails.
- Because of the phasic changes taking place due to the influence of both temperature and photoperiod, it is better to calculate photothermal units (PTU) instead of heat units for accurate prediction of flowering and maturity.
- The optimum date of sowing can be arrived by estimating PTU from climatological data.
- Number of leaves produced is less with decrease and increase in photoperiod with short and long day plants respectively.
- Tillering and plant height decrease under extreme photoperiod
- Leaf number increases with increase in temperature.
- Increase in duration of light (for long-day plants) and temperature decreases different stages of crop, especially during vegetative phase.
- A short-day plant, when exposed to short-days earlier, produces less number of leaves thus source is less.
- In rice, wheat *etc.* source is not limited but yield potential is less due to less number of storage organs.
- In sink-limited crops, amount of solar radiation in sufficient quantities is necessary during the period of formation of storage organs *i.e.*, from panicle initiation to flowering.
- In some crops, source is limited, hence solar radiation is critical during grain-filling stage, low solar radiation during this period causes large number of unfilled grains.
- Suitability of a soil to a particular crop depends on texture in addition to soil depth, depth of water table, salinity and alkalinity.
- Rice, cotton, sorghum, coriander is grown on heavy textured soils which include clay loam, silty clay loam, silty clay and clay.
- Medium textured soils like loams, silt loams, silts and sandy loams are suitable for most of the crops.

- ☆ Sandy, loamy sand, sandy loam and sandy clay are light-textured soils and these are suitable for groundnut, potato, tobacco, pearl millet and leguminous fodder crops.
- ☆ Cereals except rice are **intermediate in oxygen requirement.**
- ☆ Rice can tolerate very low level or even **complete absence of oxygen in the soil.**
- ☆ Nutrient and water uptake are reduced when the oxygen content of soil is less as it affects activity and permeability of roots.
- ☆ Volume of soil not occupied by soil particles is known as pore space.
- ☆ **Dark brown soils** indicate high organic matter and fertility.
- ☆ **Red colour** of soil shows good aeration.
- ☆ **White colour** of soil indicates accumulation of salts.
- ☆ Nutrient transformation and its availability in soils depends on pH, clay minerals, cation and anion exchange capacity.
- ☆ pH is defined as the **negative logarithm of hydrogen ion activity.**
- ☆ pH influences rate of nutrient release through its influence on decomposition, cation exchange capacity and solubility of materials.
- ☆ Important source of nitrogen and sulphur is **organic matter.**
- ☆ Decomposition of organic matter is slowest at pH below 6 and fastest between 6 to 8.
- ☆ Availability of phosphorus is high within a pH range of 6.5 to 7.5 due to higher solubility of phosphorus compounds.
- ☆ At low pH, phosphorus is precipitated as **Fe and Al phosphates**.
- ☆ At high pH, calcium phosphate is formed which is **less soluble**.
- ☆ Metallic cations such as Fe, Mn, Cu, Zn and also B precipitate at high pH, hence their availability is **less in alkaline soils**.
- ☆ Availability of K and B are influenced by pH through its effect on cation exchange capacity.
- ☆ At lower pH, H^+ ions replace K and leaching of K occurs.
- ☆ At high pH, potassium compounds are converted into nonexchangeable form and thus availability is reduced.
- ☆ Boron is leached out at **low pH.**
- ☆ Soil with pH above 8.5 is alkaline with high sodium content which deflocculates soil colloids. It results in destruction of soil structure and movement of water and air are impaired in the soil.

- ☆ At pH < 5.0, H+ ions replace Al_3^+ from the lattice of clay mineral.
- ☆ Aluminium released is hydrolysed with liberation of H+ ions. Iron behaves similarly and contributes to acidity.
- ☆ On soils with low CEC and in high rainfall regions, leaching of bases occurs resulting in reduction of pH.
- ☆ Ammonium is adsorbed on the clay complex; a portion is absorbed by the crop and around 11 per cent is lost as volatilization.
- ☆ Sine urea is soluble, leaching losses occur under lowland conditions.
- ☆ Mixing urea with soil and incubating for 24 to 48 hours results in conversion of urea into ammoniacal ion and adsorbed on complex reducing leaching losses.
- ☆ Amount of nitrogen fixed ranges from 4 to 54 per cent of total N content.
- ☆ Mineralization is faster in aerobic soils and is carried out by fungi, actinomycetes and bacteria.
- ☆ When large quantities of organic matter are added to the soils, the microorganisms utilize available nitrogen in the soils for their multiplication.
- ☆ Temporary locking up of nitrogen in microorganisms is called immobilization.
- ☆ Inorganic phosphorus is more in soil than organic phosphorus.
- ☆ Inorganic forms of soil phosphorus are **Ca-P, Fe-P and Al-P**.
- ☆ Dominant form of phosphorus in vertisols is **Ca-P.**
- ☆ Dominant form of phosphorus in alfisols is **Fe-P**.
- ☆ In Indian soils, Ca-P is around 40 to 50 per cent of total phosphorus in neutral and alkaline soils and more than 50 per cent in alkaline soils.
- ☆ At pH 2 to 5, fixation is mainly by formation of insoluble Fe and Al phosphates as Fe and Al are more in acid soils.
- ☆ pH range of 4.5 to 7.5, phosphate is fixed on clay particles by replacing two hydroxyl ions from aluminium in clay with one ion of primary orthophosphate and by forming clay phosphate linkage.
- ☆ pH range of 6 to 10, phosphate is precipitated as phosphates of calcium or magnesium as calcium and magnesium are more in alkaline range.
- ☆ Application of poultry manure decreases phosphorus availability initially, increases rapidly and attains steady state after 20 days.
- ☆ Crop production on red soils is largely limited by low availability of

phosphorus, which is attributed to adsorption of phosphate by minerals including Fe and Al oxides and Kaolinite.

- ☆ Water soluble potassium is present in the soil solution as K+ ion. It is in equilibrium with exchangeable K.
- ☆ Iron deficiency occurs in calcareous soils and soils with high phosphorus.
- ☆ Available and exchangeable iron decreases with increase in pH as it is inversely related to solubility of iron.
- ☆ Iron is **oxidized in the rhizosphere**.
- ☆ Grasses have more oxidizing power than **legumes**.
- ☆ Iron availability is reduced with oxidation.
- ☆ Iron availability is higher in **submerged soils** due to reduction of insoluble ferric compounds to soluble ferrous compounds.
- ☆ Organic matter improves availability of **iron.**
- ☆ Zinc in earth surface is mainly found as sphalerite a **Sulphide mineral.**
- ☆ In acid soils, the available manganese is up to 16 per cent of total manganese content of soils, while in alkaline soils it seldom exceeds one percent.
- ☆ Manganese availability is more in submerged soil due to reduction of insoluble manganic compounds to soluble manganous compounds.
- ☆ Actinomycetes predominates in **dry regions.**
- ☆ Tillage operations carried out throughout the year are known as **year-round tillage.**
- ☆ For wheat, soybean, pearl millet, groundnut, castor **flat levelled seed bed** is prepared.
- ☆ For maize, vegetables field has to be laid out into **ridges and furrows**.
- ☆ Sugarcane is planted in **furrows or trenches**.
- ☆ Crops like tobacco, tomato, chillies are planted with equal inter and intra-row spacing to facilitate **Two way Intercultivation**.
- ☆ **Setline planting** is adopted in Gujarat for sowing cotton and groundnut.
- ☆ In setline planting every year seed rows are in the same place, since the seed lines are set permanently at wider spacings.
- ☆ India's first digital flower auction Center is located in - **Bangalore**
- ☆ Which element is essential for hypothesis of water - **Chlorine**
- ☆ Techniques of extended shelf life of produce is - **Preservation**

- ☆ Leaves of *azolla* have the colonies of **Anabaena**"Queen of spices" is **Cardamom**
- ☆ Potassium and phosphorus content in soil should be tested in every - **3-4 years**
- ☆ Which tool used for pricking of plants is - **Shovel**
- ☆ Large part of clove imported in Indonesia is used in - **Cigarette industry**
- ☆ Pear is an example of **Temperate fruits**
- ☆ Reserpine alkaloid is found in **Sarpagandha**
- ☆ The limit of capillary rise of water in plants is **2-3 meters**
- ☆ Which type soil provides good Anchor age to tall crops **Clay loam**
- ☆ Microscope was invented by **Leeuwenhoek**
- ☆ Largest family of flowering plants is ***Orchidaceae***
- ☆ Protogyny is present in **Bajra**
- ☆ Honey bee store honey in **Cells of comb**
- ☆ What is left Behind if chloroplast is burnt **Magnesium**
- ☆ Framework elements in plants is **C, H, O**
- ☆ Plants obtained from a single homozygous and self-fertilized plants are called **Pure line**
- ☆ Causative agent of Anthrax was discovered by **R. Koch**
- ☆ Central coffee Research Institute is located at **Balehonnur (Karnataka)**
- ☆ Plant growth regulator acts as anti transpirant is **ABA**
- ☆ Who was succeeded in decoding the language of "Bee dance"**K. Frisch**
- ☆ Genetic male sterility in barley, Corn *etc.* governed by **MS gene**
- ☆ Maximum water absorption in plants take place in **Root Hair Zone**
- ☆ The sum total of all genes parents in Species is known as **Gene pool**
- ☆ Ephemeral annual plant is **Cassia tora**
- ☆ Blast resistant variety of rice is **Sabarmati**
- ☆ Chemical synthesis of gene (tyrosine *t*RNA) in lab was 1st done by **H.G. Khurana**
- ☆ Technique of isolation of nematode from soil first discovered by **N A.Cobb**
- ☆ Military dairy farm started in **1889 (Allahabad)**

- ☆ Ideal weight of egg **57 gram**
- ☆ Moisture percent in hay **15-18 per cent**
- ☆ Moisture present in silage **70 per cent**
- ☆ Burdizzo's Castrator used for **Castration**
- ☆ Trocar and conula used for **Bloat**
- ☆ Cow's milk is light Yellow or creamy colour due to **Carotene**
- ☆ White colour of cow's milk due to presence of **Casein**
- ☆ Body temperature of hen **107°F**
- ☆ Acidity per cent in milk **0.12-0.14 per cent**
- ☆ Gestation period of cow **281 days**
- ☆ Gestation period of Buffalo **310 days**
- ☆ Sugar content in milk is **Lactose**
- ☆ Glucose + glucose **Maltose**
- ☆ Glucose + Fructose **Sucrose**
- ☆ Daily salt requirement of cow **50 gram**
- ☆ Which breed of Goat known as queen of milk **Saanen**
- ☆ Best dual purpose breed of cow **Harianvi**
- ☆ Merino Sheep native from **Spain**
- ☆ Holstein friesian native from **Holland**
- ☆ Dry matter in Barseem **25-30 per cent**
- ☆ Best crop for making silage **Maize -Jowar**
- ☆ Best crop for making hay **Barseem**
- ☆ Sowing time of Barseem **Sep.-Oct.**
- ☆ Economics of agriculture is known as **Agro-economics**
- ☆ National Cooperative Union of India" was established in the year **1929**
- ☆ National Cooperative development and warehousing board was established in the year **1956**
- ☆ National Cooperative Development "Corporation" was established in the year **1963**
- ☆ National agricultural Cooperative marketing federation of India Limited was started in the year **1958**

- Central warehousing Corporation was started in the year **1957**
- Food Corporation of India was started in the year **1965**
- Vegetable oil products Act was passed in the year **1975**
- Directorate of marketing and inspection located at **Faridabad 1935**
- Food product order Act was passed in the year **1956**
- Single Window system was recommended by **Mohan Kanda**
- Money supply in economy is regulated by **RBI**
- Agro-Economic Research Centre (AERC) is located at **Sakhapatnam (AP)**
- The cost of production of the crop can be minimized by using which economic principle **Least cost combination**
- The supply of money comes under the concept **Stock concept**
- Who developed alley cropping concept **B.T. Kang**
- Forest Research Institute established at Dehradun in **1906**
- Which bioherbicide used in rice **Collego**
- Age of plants is determined by **Annual rings**
- "Handbook of Agriculture" published by **ICAR**
- Pheromone trap attract **Male Moths**
- India's lowest horsepower (HP) tractor **15 HP tractor**
- First Tractor brought in India in which year **1914**
- Which plough makes 'tf' shaped furrow **Desi plough**
- Disk plough are used when the soil is **Tough**
- Power Tiller is most suitable for the cultivation of crop **Paddy**
- What is the length of Guntur chain **66 feets**
- SRI" system of paddy was started first at **Madagascar**
- "Rice gene bank" is situated in country **Philippines**
- Parboiling in rice done to conserve **tamin B**
- Bacteria, actinomycetes, fungi and algae comes under the group of organism **Micro-flora**
- Sowing pattern used in dry land area **Broadcasting**
- Infrared thermometer used for measurement of **Crop canopy Temperature**

- ☆ Baking quality in wheat is due to **Glutin**
- ☆ Pulse which doesn't fix atmospheric nitrogen is **Rajma**
- ☆ The first fisheries institute is **CIEF, Mumbai**
- ☆ Maximum allowed biurate content of Urea is **1.5 per cent**
- ☆ Ripening hormones for fruits is **Ethylene**
- ☆ Yellow coloured fruit are rich in **tamin A**
- ☆ Potato Tuber is modified form of **Stem**
- ☆ Father of Agronomy **Pitro decrescenzi**
- ☆ Father of Agro climatology **Koppen Agro**
- ☆ Father of Meteorology **D.N. Walia**
- ☆ Father of Agriculture chemistry **Justus von Liebig**
- ☆ Father of Antibiotics **Alexander Fleming**
- ☆ Father of ATP cycle **Lippmann**
- ☆ Father of Biology **Aristotle**
- ☆ Father of Botany **Theophrastus**
- ☆ Father of Bacteriology **Leeuwenhoek**
- ☆ Father of Biochemistry **Justus von liebig**
- ☆ Father of Crop rotation **Nor fork**
- ☆ Father of Cytology **Robert Hooke**
- ☆ Father of Cytoplasmic inheritance **Carl Correns**
- ☆ Father of Cooperative Movement in India **N. Nicholson**
- ☆ Father of DNA fingerprinting technique **Alec Jeffrey**
- ☆ Father of Economic Ecology **M.S. Swaminathan**
- ☆ Father of Extension education A. **Semen and Knapp (USA) Leagans (India)**
- ☆ Father of Experimental Genetics **Thomas hunt Morgan**
- ☆ Father of Field plot experiment **J. B. Boussingault**
- ☆ Father of Fruit and vegetable preservation **Nicolas appert**
- ☆ Father of Genetics **Gregor John Mendel**
- ☆ Father of Genetic engineering **Paul Berg**
- ☆ Father of Green Revolution **Dr N E Borlaug**

- ☆ Father of Green revolution in India **MS Swaminathan**
- ☆ Father of Golden revolution in India **Dr K.L. Chadda**
- ☆ Father of Golden rice **Dr. Ingo Potrykus**
- ☆ Father of Hybrid rice **Yuan long ping**
- ☆ Father of Hybrid cotton **C.T. Patel**
- ☆ Father of Indian plant pathology **E.J. Butler**
- ☆ Father of Indian mycology **E.J. Butler**
- ☆ Father of Indian Ecology **R. Mishra**
- ☆ Father of Indian Plant breeding **Dr.B.P Pal**
- ☆ Father of Indian Remote Sensing **Dr PR Pishroty**
- ☆ Father of Immunology **Edward Jenner**
- ☆ Father of Medicinal biotechnology **Robert Koch**
- ☆ Father of Modern genetics **T. H. Morgan**
- ☆ Father of Mutation theory **Hugo de vries**
- ☆ Father of Modern botany **Linnaeus/Bauchin**
- ☆ Father of Modern cytology **Swanson**
- ☆ Father of Nematology **N.A. Cobb**
- ☆ Father of Nitrogen fixation **Winogradsky**
- ☆ Father of Ornamental gardening **M.S. Randhawa**
- ☆ Father of Plant pathology **Anton De Bary**
- ☆ Father of Plant physiology **Stephen hales**
- ☆ Father of Pedology **V. V. Dokuchalev**
- ☆ Father of Parasitology **F.Platter**
- ☆ Father of Plant tissue culture **G.Haberlandt**
- ☆ Father of Plant anatomy **Grew**
- ☆ Father of Polygenic inheritance **Kolreuter**
- ☆ Father of Pure culture techniques **Oscar Brefield**
- ☆ Father of Sociology **Auguste Compte**
- ☆ Which Chemicals are commonly used for preservatives in India **Sodium Benzoate and Potassium Metabisulphate**

- ☆ Prevention of micro organism entry in fruit and vegetable called as **Asepsis**
- ☆ Maximum solubility of salt in a solution is **26.5 per cent**
- ☆ Grape yeast; *Saccharomyces cerevisiae* and *Saccharomyces ellicidae* are used for **Wine production**
- ☆ Which chemical used for preserving colourless product **Potassium Metabisulphate**
- ☆ Which chemical used for preserving coloured product **Sodium Benzoate**
- ☆ Which type containers are used for Canning of coloured grape in caning **Enamel**
- ☆ Strength of Brine is measured by using **Salinometer**
- ☆ which food colour banned in India **Acid Magnet-II**
- ☆ Which vegetable colour banned **Gamboge**
- ☆ Which hormone used for antioxidant **Ascorbic Acid**
- ☆ Which enzymes used for clarifying **Beverages Pectinase**
- ☆ Papain enzymes are used for **Tenderizing meat**
- ☆ Which enzyme used for producing cheese and curd **Renin**
- ☆ Which chemical work as meat curing **Sodium nitrate**
- ☆ Semi perishable fruit and vegetable are **Apple and Potato**
- ☆ Killing of all viable microbes in a food product known as **Sterilization**
- ☆ Prevention by Irradiation is called as **Cold Sterilization**
- ☆ Central Food Technology Research Institute (CFTRI) located in **Mysore, Karnataka**
- ☆ Hydrogenated fat is common adulterant in **Butter and ghee**
- ☆ Nutritionally perfect food is **Egg**
- ☆ The best way of preserving produce is **Freezing**
- ☆ The main function of carbohydrate is **Supply energy**
- ☆ Who is the Father of food nutrition **Antoim levoiser**
- ☆ Father of Canning **Nicholas apart**
- ☆ Main aim of curing **To heal wounds**
- ☆ Main aim of blanching **To inactivate enzymatic activity**
- ☆ Class II preservatives **BA and KMS**

- ☆ Class 1 preservatives **Salt, sugar, spices**
- ☆ KMS used for preservation of **Colour less products**
- ☆ Perosis is caused by **Mn deficiency**
- ☆ Poultry raised for meat purpose **Boiler**
- ☆ "Strengthening infrastructure for quality and clean milk production scheme" started in **October 2003**
- ☆ Milk and milk product order passed in **1992**
- ☆ Key Village scheme started in **1950-52**
- ☆ Ideal weight of egg **57 gram**
- ☆ Moisture present in silage **70 per cent**
- ☆ Burdizzo's Castrator used for: **Castration**
- ☆ Moisture percent in hay **15-18 per cent**
- ☆ Cow's milk is light Yellow or creamy colour due to **Carotene**
- ☆ Trocar and conula used for **Bloat**
- ☆ Acidity per cent in milk **0.12-0.14 per cent**
- ☆ Body temperature of hen **107°F**
- ☆ Gestation period of Buffalo **310 days**
- ☆ Gestation period of cow **281 days**
- ☆ White colour of cow's milk due to presence of **Casein**
- ☆ Daily salt requirement of cow **50 gram**
- ☆ Sugar content in milk is **Lactose**
- ☆ Which breed of Goat known as queen of milk **Saanen**
- ☆ Best dual purpose breed of cow **Harianvi**
- ☆ Holstein friesian native from **Holland**
- ☆ Merino Sheep native from **Spain**
- ☆ Dry matter in Barseem **25-30 per cent**
- ☆ Best crop for making hay **Barseem**
- ☆ Best crop for making silage **Maize –Jowar**
- ☆ Cracked feet in poultry is caused by **Deficiency of Biotin**
- ☆ Sowing time of Barseem **Sep.-Oct.**
- ☆ Temperature for artificial vagina **42°**

- ☆ Anthrax is caused by **Bacteria**
- ☆ Removal of tail in sheep **Docking**
- ☆ Highest and sweetest milk producing cow breed in India **Sahiwal**
- ☆ Fat content in double tonned milk is **1.5 per cent**
- ☆ Parakertosis is caused by deficiency of **Zinc**
- ☆ Colour of egg yolk is due to: **Xanthophyll**
- ☆ Tillage improves: **Infiltration, water holding capacity**
- ☆ Tillage operations are grouped into two types such as: **Primary and secondary tillage**
- ☆ Operation carried out from the harvest of the crop to before sowing is called as: **Preparatory tillage**
- ☆ Disc harrowing and blade harrowing are: **Secondary tillage operations**
- ☆ Primary tillage practices are: **Ploughing, planting and harrowing**
- ☆ Weeding is the practice of: **Secondary tillage operation**
- ☆ Hoeing before germination is called as: **Blind tillage**
- ☆ The sowing method that ensures uniform plant population is: **Dibbling method**
- ☆ Deep tillage is preferred under: **Long duration, deep rooted crops and dry land condition**
- ☆ Chisel plough mostly used for: **To break the subsoil compaction**
- ☆ Depth of plough depends on: **Soil moisture and soil texture**
- ☆ The optimum range of moisture for effective plough is: **25 -50 per cent**
- ☆ Bulk density of tilled soil is: **Higher**
- ☆ Bulk density of untilled soil is: **Less than tilled soil**
- ☆ Summer deep ploughing improves: **Soil Structure**
- ☆ Soil considered as good structure with: **Granular and crumbly**
- ☆ The capillary and non-capillary pores are about equal when the soil is called as: **Good tilth**
- ☆ Which Soils size are considered favorable for crop growth: **1-5 mm (crumby structure)**
- ☆ The critical level of oxygen for better root growth is about: **1-2 per cent**
- ☆ The critical limit of CO_2 for most species is between: **10 - 20 per cent.**

- ☆ The depth of the primary tillage varies from: **10-30 cm.**
- ☆ When soil tilled from 25 – 45 cm is called as: **Deep tillage**
- ☆ Black soil requires the number of plough is: **3-5 times**
- ☆ Summer shower is used for tillage operation is called as: **Summer ploughing**
- ☆ Rotary plough is used for: **Cut the light soil**
- ☆ Which plough is used for deep ploughing in grass field: **Disc plough**
- ☆ Minimum tillage is: **Raising of crop with least tillage practices**
- ☆ Zero tillage was first adopted by: **G. B. Triplet**
- ☆ Zero tillage was first used in: **USA**
- ☆ Zero tillage requires the minimum ploughing of: **One**
- ☆ Zero tillage is: **Extreme form of minimum tillage**
- ☆ Which herbicides are used in zero tillage: **Paraquat and Diquat**
- ☆ Zero tillage method is highly effective under: **Sloppy lands**
- ☆ No tillage method has been successfully adopted for planting under: **Grasses and Legumes**
- ☆ Country plough angle of about is: **120° to 135°**
- ☆ Which type furrow is formed by Country plough: **V' shaped furrow**
- ☆ The depth of the furrow is usually: **10-20 cm and the width varies between 13 and 18 cm**
- ☆ The country ploughing is always started from: **Right side of the field**
- ☆ Ploughing should be done across: **The slope to reduce loss of top soil**
- ☆ The 'knife coulter' is used in: **Grasslands**
- ☆ The 'disc coulter' is used in: **Fields infested with weeds**
- ☆ In mould ploughing is started from the center of the plot and is turned round and round always towards the mould board side. Lastly a ridge is left in the center of each plot is called as: **Gathering' method**
- ☆ In mould ploughing is started at the two opposite edges of the plot. Ploughing is continued inward and completed with a furrow in the center is called as: **Splitting method**
- ☆ Mould Plough is turned always towards the: **Landside**
- ☆ Disc ploughs have steel discs of: **50-90 cm diameter**

- Cultivators are otherwise called as: **Tillers (India)**
- The chisel plough was developed at: **Tamil Nadu Agricultural University, Coimbatore**
- Star weeder is used for: **Weeding in Dry Land**
- Wooden Blanks are used for: **Micro levelling/Land levelling**
- Basin lister is used for: **To make basins to prevent run off**
- Which cultivator is used in both harvesting of groundnut and in stubble mulch: **Sweep cultivator**
- Dibbling is a method of: **Sowing**
- Shifting cultivation is also known as: **Puham or Jhumming**
- In irrigated area, soil aggregated size should be: **>5 mm**
- In dry land areas, soil aggregated size should be: **1-2 mm**
- Objective of Puddling is: **Reduce deep percolation losses, water and weed control**
- Other name of mulch tillage is: **Fallow and clean tillage**
- Reduced tillage is also known as: **Conservation tillage**
- Which weeder is used for weeding in closer spacing line sown crops: **Star weeder**
- Minimum tillage operations are: **Row zone, plough plant and wheel tract tillage**
- Row zone tillage comes under: **Secondary tillage**
- Conservation tillage is also known as: **Stubble mulch tillage, eco fallow reduced tillage**
- In which soils are used peg type weeder: **Loamy and clay soils**
- What is the size of fine tilth aggregates: **0 – 5 mm**
- Which plough is used for drainage: **Mole plough**
- The minimum seed rate required for the method of sowing is: **Dibbling method**
- Puddling is an essential practice for: **Transplanted rice**
- Minimum tillage system is particularly advantageous in: **Heavy clay and clay loam soils**
- Plough plant system method is most suitable on: **sandy loam soils**

- Which element required small amount to crop growth- **molybdenum** (Molybdenum nutrients has amphoteric nature (both react at acid and base condition)
- During early stage rice crop taken at ammonia form and late stage rice crop taken **nitrite form.**
- Metal nutrients are (Cation) - **K, Ca, Mg, Mn, Fe, Co, Zn, Nickel form**
- Non-Metal (Anion) nutrients are- **C,H,O,N,P,S,Cl form**.
- Highly mobile nutrient taken at early stage (**Nitrogen, Sulphur, Boron**)
- Plant tissue contain **carbon - 45 per cent, oxygen-43 per cent, hydrogen-6 per cent**
- Carbohydrate ratio **1:2:1**
- Nitrogen elements only taken **both anion and cation form**.
- In acid soil nitrification process slow because plant will take through **ammoniacal form**
- All the micro nutrients available decrease in more pH (expect Molybdenum available in alkaline condition)
- In compact soil/dry soil condition ethylene production increase. When ethylene production and micro hormone increase in soil generally **SHUT DOWN** of shoot and root growth and cellular activity.
- High CEC- **low nutrient held in soil and decrease mobility and uptake of nutrients.**
- Nitrite –**more susceptible to leaching**
- Ammonia - **most resistant to leaching**
- Sulphur nutrients only **mobile in the soil**
- In flood condition increase solubility of **Iron and Manganese nutrients** Example Black Gram crop flooded condition Mn, Fe toxicity symptom will appear in leaves
- **Poor aeration condition soil –** H_2S formation (one-week flood or water stagnation in field are produce ethylene and leaf senescence appear)
- In **dry condition** Iron element taken in the form of **Ferric form**
- In **water logged condition** Iron element taken in the form of Ferrous form (during rainy period leaves are dark green colour due to Ferric is converted into Ferrous form and thunder storm are produce nitrogen to the crop)
- Ideal temperature for available nutrients- **25-35°C**

- Total chlorophyll of leaf measured by **Acetone method, TMOS, SPAD meter**
- In alkaline soil phosphate fertilizer are recommended - **Mono ammonium phosphate is more stable**

  Potassium nutrient available in Acid soil condition (If application of Liming materials increase the availability of potassium nutrients) Three mechanisms involving nutrient uptake by plant from soil medium

- Root interception – Ca nutrient taken
- Mass flow mechanism- N, Mg, S.
- Diffusion mechanism- P, K.
- Diffusion process is very slow compare than **root interception and mass flow**
- The CEC of root in monocot crop range **10-30Meg/100g of root**
- The CEC of root in dicot crop range **40-100Meg/100g of root**
- In monocot crop need **- K^{+}, Ca^{+}**
- In Dicot crop needs- **Ca^{++}, Mg^{++}**
- **Apo plastic** is defined - minerals and nutrients along with water moves from cell to cell through spaces between cell wall by diffusion process (normally in non living plant part
- **Symplastic** is defined - minerals and nutrients enters the cytoplasm of epidermis move across the cytoplasm of cortex, endodermis of pericycle through plasmodesmata finally reach the xylem (normally in living plant part).
- **Hidden hunger** or subclinical disorder common in Fruit and plantation crops
- Nitrogen: Sulphur in soil ratio between- **10:15**
- Nitrogen: potassium in soil ratio between- **1.2-2.2**
- In cereals Nitrogen: potassium in soil ratio between- **1:1.2**
- In deep tillage practices build up **more Ca, P nutrient**
- In drought condition slow down reduce of nutrients to root and create deficiency of **N, K, Cu, B**
- **Autochthonous or Native microbes** means indigenous, which are characteristic of the particular soil and which many be expected always to be found their *e.g.,* Arthrobacter

- **Zymogenous or fermentative organism**- require an external source for normal population in soil is low. when specific substrates are added to soil the population is increased *e.g., Pseudomonas, Bacillus*
- Aerobic means- **need O_2 for growth**
- Anaerobic means- **not require O_2 for growth**
- Facultative anaerobic – **live in the presence or absence of O_2**
- Biotrophs means- **Carbon from plant to exchange for microbes**
- Examples of photoautotrophs – **Algae** (*Cyanobacteria*)
- **Growth curve** of microbial population is arithmetic means steady/ systemic growth rate and logarithmic curve means suddenly change of growth
- Optimum temperature for microorganism population - **25-35°C**
- Normal soil pH for survey of microorganism population- **6.2-7.0**
- *Acidophilus* means to survey on pH from- **4.0-6.0**
- *Protazae* in dry soil - **Cyst (dormant) condition**
- Algae indicator for **Air pollution**